August Gattinger

Tennessee Flora

With Special Reference to The Flora of Nashville

August Gattinger

Tennessee Flora
With Special Reference to The Flora of Nashville

ISBN/EAN: 9783337270513

Printed in Europe, USA, Canada, Australia, Japan

Cover: Foto ©berggeist007 / pixelio.de

More available books at **www.hansebooks.com**

THE

TENNESSEE FLORA;

WITH SPECIAL REFERENCE TO

THE FLORA OF NASHVILLE

PHÆNOGAMS AND VASCULAR CRYPTOGAMS.

BY

DR. AUGUST GATTINGER,

MEMBER OF AMERICAN ASSOCIATION FOR ADVANCEMENT
OF SCIENCE.

PUBLISHED BY THE AUTHOR.
NASHVILLE, TENNESSEE.
1887.

PRINTED BY
CARLON & HOLLENBECK,
INDIANAPOLIS, IND.

PREFACE.

Desiring to promote the study of botany in the educational institutions of the State, and to awaken an interest in the exploration of the Flora of Tennesseee, I tender this small volume to the friends and promoters of scientific pursuits.

The work is original, being founded upon a botanical collection made exclusively by myself, during thirty-eight years' residence in this State.

I am yet in possession of specimens collected in 1849, when I first took up my residence in East Tennessee as a practicing physician. Placed, as I was in those early days, amid unfamiliar modes of life, with no access to intellectual resources, without information about the condition and advance of scientific affairs in this country, my botanical progress could for many years be no other than tedious and slow; but I kept up a pursuit, which since early school-years had been to me a source of pleasure and consolation.

After my removal to Nashville, in 1864, I paid special attention to the exploration of the vicinity of Nashville and the adjoining counties. Travel by railroads made it possible to make frequent short visits to distant points, without too great infringement on professional duties.

Although in an educational center, filled with display of refinement, I soon perceived that I had to rely upon my own resources, if I would attempt to expand my botanical efforts beyond the limits of personal gratification. For the want of such blessed leisure as would be needed to assure success, I never expected to publish on the Flora, considering the results of my investigations too insignificant. That I have now prepared this paper is purely contingent upon the meeting of the American Association for the Advancement of Science, in Nashville, August, 1877.

At that occasion I had the good fortune of making acquaintance with some well-known Eastern botanists, who, with very obliging politeness, reviewed my collections, and assured me that a survey of the unexplored region of Tennessee would be appreciated. At their instance I continued, with all care and

pains possible, to make the work true and reliable, with the hope
and solicitude to make this insignificant, but to me only possible
one contribution, to American science.

I am under lasting obligations to William N. Canby, Esq.,
of Wilmington, Del., Prof. J. W. Chickering, Jun., and Prof.
Lester F. Ward, both of Washington, D. C., for their advice and
the attention they paid to me at the Nashville meeting. From
that time on I also enjoyed the privilege of submitting critical
specimens to Dr. Asa Gray, of Cambridge, for his decision. The
late Dr. George Engelmann, of St. Louis, Dr. A. W. Chapman,
of Apalachicola, Fla., and Dr. George Vasey, botanist of the
Department of Agriculture, have relieved me of many doubts
and supplied me with a great number of authentic specimens. I
shall ever gratefully remember Dr. Engelmann, and express to
the other gentlemen my sincere thanks. Acknowledgments are
also due to many active botanists in distant parts of the Union,
for their readiness and promptness in exchanging plants and
opinions.

I have no knowledge of authentic published records bearing
on the Flora of Tennessee, except an article contributed by Prof.
J. W. Chickering to the *Botanical Gazette*, December, 1880, enti-
tled "A Summer on Roane Mountain." In a number of *Sullivant's
Journal*, of 1841, I find a sketch of a botanical tour through the
Alleghanies and on Roane Mountain, by Dr. Asa Gray.

It is much to be regretted that Dr. Rugel, who about thirty
years ago resided in the vicinity of Greenville and made valuable collections and discoveries in that vicinity, and the mountains of East Tennessee and North Carolina, died without leaving
a record of his work. His collections came in the possession of
Mr. Shuttleworth, of England. Senecio Rugelia Gray, Plantago
Rugelii Decaisne, Siphonychia Rugelii Chapm. commemorate his
name.

Some species and stations which fell not under my personal
observation, are quoted on good authority or credited to the col-
lector.

For description of species I refer to Dr. Gray's Manual and
Dr. Chapman's Flora of the Southern States.

I am fully aware of the incompleteness of the work, but
enough is now done to give a satisfactory estimate of our Flora,
and nothing short of publication can near its completion.

Pretending to no other merit than one, due to a persevering
effort to illustrate the distribution of the American Flora over
the territory of the State of Tennessee, and the accidental dis-

covery of some new species, I hope for a kind reception by those
for whose benefit it is intended.

My botanical friends will appreciate the difficulties I had to
encounter, and I solicit contributions and corrections from those
who are favorably inclined towards its improvement.

Parting I embrace the opportunity to urge a higher appreci-
ation of the study of Botany, and to sum up the present standing
of the science, and the advantages it enjoys in other States.

American botany has made rapid advances within the last
decade. The earlier periods passed over in engagement with sys-
tematic work, collecting and systematizing phænogams and the
higher cryptogams. The general survey being very far advanced
and nearing completeness, by degrees a state of maturity for
studies of a higher order has been entered upon, which demands
greater proficiency in analysis and dexterity in the use of the
microscope.

The intricate study of the life-history of plants of the lowest
orders engages now the attention of our more advanced botanists.
Detail in physiology and morphology and original work is also
fairly attempted. A number of our progressive American universities have attached laboratories to the botanical lecture-rooms,
and provided them with the necessary outfits, as powerful auxili-
aries to the study of botany. Harvard and Cornell since 1872,
more recently the universities of Pennsylvania and Michigan,
Iowa Agricultural College, Wabash College, Purdue University,
the universities of Wisconsin and Nebraska, Shaw School of
Botany at St. Louis, etc.

Why is it, that in our more than centennial State, so little
has been done for the improvement of natural knowledge? What
object of teaching can conduce more to the material welfare and
and progress of the citizen than a practical information how to
disclose the concealed wealth, to collect and utilize wasting ener-
gies, to draw from the soil maintenance of life and means of com-
fort, without impairing its productiveness? And what can more
than the improvement of natural knowledge enlargen and ele-
vate the intellectual ethics of man, than the ever growing con-
ception of a definite and uninfringible order of nature? What
can add more to his personal dignity, inspire him with more self-
reliance, than the certainty of possessing means to test and verify
his conceptions, by bringing them in contact with Nature herself,
by experiment and observation?

Having inadvertently diverted from the proposed plan of
this address, I shall avoid to make further reflections for the
same caution with which the astray botanist avoids the treacher-

ous briers. Should I have succeeded to bring you to notice that
rambling through field and forest after plants, implies a higher
purpose than the pleasure of analyzing and adding them to the
collection, then I would feel like the aberrant botanist, who,
aberrant, made a precious discovery. Remember of all, that it is
not for every one to penetrate into the depths of a science, but
that a plain and correct knowledge of the leading principles in
botanical science is attainable and useful to all.

" Ingenuas didicisse fideliter artes
  Emollit mores nec sinit esse feros."

A. GATTINGER.

Nashvilee, Tenn., Feb. 3, 1887.

GENERAL ASPECT OF THE FLORA.

The boundaries of Tennessee are embraced within the great Atlantic forest region. The whole of it was in its virgin state, a congeries of varied woodlands, being in the lowlands of dense and massive growth, filled with pathless jungles of cane and shrub, or, away from the watercourses, on the uplands, reduced to open and airy groves, the barrens. Here a dense sward covers the ground and herbaceous growth prevails. Mountain forests have always been of greater uniformity in distribution of timber.

Nearly one-third of the entire area is now reduced to fields or occupied by buildings or roads. Canebrakes have well nigh disappeared, and the forest is in all accessible regions depleted of valuable timber.

Immigration of foreign and retirement of native species continually modify the aboriginal flora a..d tend to weaken characteristics due to presence of peculiar plantforms, or collocation of species, by the intricacies of mutual predilection and adaptation to surroundings.

Such areas, which differ amongst themselves conspicuously in such properties, admit of the establishment of natural floral arrondisements.

Differences of elevation, diversity in elementary constitution of the soil, and inequality in distribution of atmospheric humidity are, in our territory, sufficiently potent to mark out four distinct regions.

I. The high crests of the Alleghany Mountains, formed of gneiss or mica-shists, with an elevation from 4,000 to 6,600 feet. Subalpine region.

II. The western slopes of the Alleghanies and their outlying spurs, and the Cumberland Mountains. Sandstones and slates. Mountain-flora. Elevation 2–4,000 feet.

III. Valley-flora, the lower division of which is coëxtensive with the limestones (silurian) of East and Middle Tennessee. Elevation 350–500 feet. The upper division or highlands has siliceous and argillaceous soils, sometimes limestones of the sub-carboniferous formation. Elevation about 1,000–1,200 feet.

The former division is characterized through its cedar glades; the latter is the region of the oak-barrens.

IV. West Tennessee, situated between two powerful rivers, with much level or only gently undulating surface, owes its peculiarities to the abundance of swampy lands and predominantly argillaceous soils, in connection with a more humid atmosphere.

I. SUBALPINE REGION.

The dividing line between the States of North Carolina and Tennessee passes over and along the crest of the highest ridges and peaks, known as the Unaka, Great Smoky, Bald and Frog mountains. Their average elevation is about 5,000 feet, but about twenty-two summits are 6,000 feet or more. The highest stretch lies between French Broad and Little Tennessee rivers, with fifty-five high points, eighteen of which are over 6,000 feet. Clingman's Dome, by a few feet the highest, rises to the very respectable altitude of 6,660 feet above tidewater, according to the measurements of Prof. Arnold Guiot, of Princeton, N. J. (Vide *Am. Jour. Science*, Sept. 1857 and Nov. 1860). Geologically they consist of Huronian shists and gneisses, and in some spots of Laurentian granites.

Not one of these high crests presents a bleak crag, bare of vegetation, nor is there a timber-line. Some are evenly timbered throughout, others support only a scattered and stunted arboreal growth, and some bear only a low shrubby or herbaceous vegetation. The absence of timber on the so-called "Balds" is perhaps due to waves of excessive cold; such, at least, seem the naked trunks looming up here and there, to suggest. There are generally groups of red oak, striped or mountain maple, mountain ash and chestnut, with open spaces between. They have a stunted and gnarled appearance, their sprawling limbs often but few feet from the ground. Chestnuts of the summit flower from three to four weeks later than those in the neighboring valleys.

Upon these lofty retreats dwells a limited number of species peculiar to those regions. Many more are denizens of the common flora of more northern latitudes, but are not found in the intervening lowlands. The abundance of evergreen Rhododendrons, exuberant flowering ericaceous and liliaceous plants produce a floral scenery of unsurpassing beauty. From the altitude of the region, as well as from the physiognomy of the vegetation, we may call it "subalpine."

To here belongs *Abies Fraseri* Pursh. and *Rhododendron Catawbiense* Michx. *Betula lutea* Michx. grows to moderate size,

and Hamamelis Virginica I found as a chunky tree, measuring
fifty inches in circumference of trunk. *Alnus viridis* DC., *Menziesia globularis* Salisb., *Salix humilis* Marsh., *Vaccinium corymbosum* L. var. *pallidum* Gray, *Vaccinium erythrocarpum* Michx.,
Leiophyllum buxifolium Ell., *Ribes rotundifolium* Michx., make
up the small shrubbery. Of herbaceous plants I would mention
lilium superbum L., justly called so, every one would concede,
who had traversed at the beginning of summer those waving
mountain-savannahs, in which it abounds. It attains a height of
six feet, with pyramidal racemes bearing as many as twenty-five
blossoms. *Melanthium Virginicum* L., a worthy companion,
spreads its enormous panicles at the same time. Other interesting objects are *Melanthium parviflorum* Gray, *Stenanthium angustifolium* Gray, *Streptopus roseus* Michx., *Clintonia umbellata* Jarv.,
Lilium Grayi Wats., *Convallaria majalis* L., *Heracleum lanatum*
Michx., *Angelica Curtisii* Buckl., Rubus odoratus L., Galium
latifolium L., *Hypericum graveolens* Buckl., *Danthonia compressa*
Aust., *Muhlenbergia Willdenovii* Trin., *Deyeuxia Nuttalliana* Vasey,
Carex trisperma Dew., *C. æstivalis* M. A. Curtis, *C., juncea* Willd.,
Triestum palustre L., *Deschampsia flexuosa* Beauv., *Agrostis
canina* var. *rupestris* Chapm., *Aspidium spinulosum* Swarz, *Dick-
sonia punctilobula* Kunze, *Lycopodium Selago* L. All the above
abound in the Balds. More scattered over the region we find:
Solidago glomerata Michx., infested with *Cuscuta rostrata* Shuttel-
worth, *Aconitum reclinatum* Gray, *Delphinium exaltatum* Ait.,
Trautvetteria palmata Fish. & Mayer, *Solidago spithamea* M. A.
Curtis, *S. monticola* Torr. & Gray, *Houstonia serpyllifolia* Michx.,
Circæa alpina L., *Oxalis Acetosella* L., *Vaccinium hirsutum* Buckl.,
Sedum Rhodiola DC., *Saxifraga leucanthemifolia* Michx., *S. Car-
eyana* Gray, *Paronychia argyrocoma* Nutt., *Nabalus Roanensis*
Chick., *Viola renifolia* Gray, *Cardamine Clematitis* Shuttelwo.,
Parnassia asarifolia Vent., *Krigia montana* Nutt., *Geum genicu-
latum* Michx, *G. radiatum* Michx., *Arenaria glabra* Michx.

Descending into the second region, we find *Gaylussacia brachycera* Gray, and *Vaccinium hirsutum* Buckl. forming the common
undergrowth (on Big Frog Mountain), *Gaultheria procumbens* G.,
Rhododendron maximum L., *R. calendulaceum* Torr. and *R. viscosum* Torr., *Clethra acuminata*, Michx. This is the finest pine
region of the State. *Pinus Strobus* L. and *Tsuga Canadensis*
Carr. grow to the largest dimensions. *Pinus pungens* Michx.,
intermixed with *P. rigida* Mill., predominates in several districts
in the Big Smoky Mts., while *P. mitis* Michx. and P. inops are
more at home on the lower spurs.

Moist and shady ravines, in which humus accumulates, suits

the magnolias. *Magnolia Fraseri* Walt. and *M. macrophylla* Michx. are very conspicuous from the unwonted size and fresh green color of their foliage. *Ilex opaca* Ait., *Nyssa Caroliniana* Michx., and *Tsuga Canadensis* Barirèr follow every rill and run along which long lines of *Leucothœ Castesbaei* Gray form impassable barriers. Remote mountain glens, where one has to push his way through branchy hydrangeas or prickly azalias, or to crawl through the zigzag limbs of big laurels, ought to be approached with caution. It is irksome and dangerous to be entrapped in such labyrinths. Sunnier and higher positions are chosen by *Pyrularia oleifera* Gray, *Buckleya distichophylla*, Torr., *Calycanthus glaucus* Willd., *Ilex monticola*, Gray and *Corylus rostrata* Ait. Of climbers we note *Aristolochia Sypho* L'Her., *Decumaria barbara* L., *Celastrus scandens* L. Out of a rich display of herbaceous plants I would select *Lysimachia Fraseri* Duby, *Oenothera glauca* Michx., *Diphylleia cymosa* Michx., *Adlumia cirrhosa* Raf., *Dicentra eximia* DC., *Draba ramosissima* Desv., *Viola Canadensis* L., *Ascyrum hypericoides* L., *Baptisia tinctoria* R. Br., *B. alba* R. Brown, *Thermopsis fraxinifolia* M A. Curtis, *Waldsteinia fragarioides* Tratt., *Potentilla tridentata* Ait., *Saxifraga erosa* Pursh., *Sedum Nerii* Gray, *Chrysogonum Virginianum* L., *Helianthus lævigatus* Torr. & Gray, *Campanula divaricata* Michx., *Galax aphylla* L., *Melampyrum Americanum* Michx., *Pycnanthemum montanum* Michx., *Monarda didyma* L., *Gentiana quinqueflora* Lam.

Creeks and brooklets have their rocky bottoms lined with the curious, mosslike *Podostemon abrotanoides* Michx., which disappears whenever the current's speed is checked and the channels deepen.

Another range of mountain flora we find in the Cumberland mountains. Selecting the Lookout near Chattanooga for a type, we find its summit wooded with *Quercus Prinus* L., *Q. rubra* L., *Q. alba* L., *Q. obtusiloba* Michx. and *Q. nigra* L, *Pinus inops* Ait., *P. Tæda* L., *P. mitis* Michx., *Betula lutea* L., *Gleditschia triacanthos* L., *Robinia Pseudacacia* L., several Caryas and *C. microcarpa* Nutt. among them. Of shrubs: *Robinia hispida* L., *Diervilla sessilifolia* Buckl., *Ilex mollis* Gray, *Stuartia pentagyna* L'Her., *Hydrangea radiata* Walt., and again (but very rare) *Buckleya distichophylla* Torr., *Nemopanthes Canadensis* DC., and in a swamp *Dirca palustris* L. Of herbaceous plants: *Utricularia gibba* L., *Iuncus Canadensis* J. Gay, and *Arundinaria tecta* Muhl. On flat rocks: *Diamorpha pusilla* Nutt., *Fimbristylis capillaris* Gray, *Krigia Virginica* Willd., *Arenaria glabra* Michx. On the cliffs of the crest: *Stipa avenacea* L., *Silene rotundifolia* Nutt., *Linaria Canadensis* L., *Campanula divaricata*

Michx., *Thalictrum clavatum* DC. Near the base of the mountain, on limestone ledges, *Gatesia lætevirens* Gray, *Callicarpa Americana* L., *Triosteum perfoliatum*, L., *Silphium brachiatum* Gattinger. The Cumberlands excel the Alleghanies in a greater variety of ferns. Besides all species of the latter, we also find here *Asplenium Bradleyi* Eat., *A. pinnatifidum* Nutt., *Lygodium palmatum* Swartz., *Scolopendrium vulgare* Smith, and *Trichomanes radicans* Swartz.

The third division embraces the valley of East Tennessee and the entire area of Middle Tennessee. Contour of surface and geological structure result in East Tennessee from the combined processes of folding and erosion, whereby heterogeneous strata are placed in juxtaposition, the whole valley being an often-repeated series of synclinals and anticlinas of calcarious and siliceous rocks, while in Middle Tennessee erosion alone had been at play.

A great fault connected with the upheaval of the Pine and Crab-orchard mountains, and in a line south of it, an eroded anticlinal, the Sequatchee valley, designate in the Cumberland mountain region the western terminus of those convulsions which involve the problem of the stratography of the Alleghanies in so great difficulties. West of this line spread out the horizontal strata of the Cumberland table-land, which terminates with an abrupt descent of about one thousand feet upon the highlands of Middle Tennessee. These in turn overreach and encircle the floor of the basin of Middle Tennessee by five to six hundred feet, either in a bluff or through a gradual descent.

The succession of strata is normal throughout: uppermost subcarboniferous limestone and chert, followed by the Devonian shale, lastly the lower silurian.

Increase in annual range of temperature and greater dryness of air, as compared with the former regions, cause the mountain flora to disappear and to yield to other designs in nature's garb. A close botanical inquiry into the array of species soon discloses the fact that different assemblies of species congregate in the limestone and argillaceo-siliceous region. The former includes the glades, the latter the barrens of Middle Tennessee.

Glades are thinly wooded, unarable lands, with shallow soils, fit only for pastures. They ought to remain in their natural state, undisturbed by cultivation. To clear them is to convert them into deserts. In some parts they are exclusively occupied by the cedar, with a small percentage of deciduous trees intermingled. In other places prevails the Ohio buckeye, (*Æsculus*

flava Willd.), honey locust (*Gleditschia triacanthos* L.), hack-
berry (*Celtis Mississippiensis* Bosc.), some hickories (*Carya alba*
Nutt., *C. tomentosa* Ntt., *C. porcina* Nutt.), shingle oak (*Quercus imbricara* L.), yellow chestnut oak (*Q. Muhlenbergii* Englm.),
post oak (*O. obtusiloba* Michx.), hop-hornbeam (*Ostrya Virginica* L.), winged elm (*Ulmus alata* Michx.), buckthorn (*Frangula
Caroliniana* Gray), persimmon (*Diospyros Virginiana* L.), red
plum (*Prunus Americana* Marshall), Chickasaw plum (*P. Chickasaw* Michx.) Of shrubs prominently: *Forestiera ligustrina*
Poir., *Rhus aromatica* Ait, *Ptelea trifoliata* L., *Aralia spinosa*
L., several hawthorns: *Crataegus Crus Galli* L., *C. cordata* Ait.,
C. tomentosa L., var. *pyrifolia* Gray, and var. *punctata* Gray,
Bumelia lycioides Gært., *Symphoricarpus vulgaris* Michx., *Hyper-
icum aureum* Bart.

The cedar barrens effect an obvious and pleasing contrast in
the feature of a landscape, especially in regions where, by absence
of streams or prominent landmarks, diversity in grouping, habitus and coloring of the arboreus growth must relieve a weari-
some monotony. Middle Tennessee is, from periodic excessive
dryness of the atmosphere, absolutely incongenial to every other
species of our native conifers.

The somber tint of the cedar delineates a cedar barren from
its surroundings at a distance and serves within its environs,
wherever openings occur, to bring out with dazzling vividness
the crumbling limestone flats, overspread with the rosy Sedum
pulchellum and carmine-flowered Talinum, or the golden stars
of the Opuntia Raffinesquii.

The botanical interest in these cedar glades varies from a
delightful surprise in the survey of an unparalleled number of
rare and interesting plants upon small tracts, to a painful disap-
pointment over a fruitless ramble through long stretches.

Depressions, where the coherent and slightly scooped limestone banks secure a continuance of moisture, and where small
springs come to the surface, represent the garden spots of the
wilderness. Wherever again the ground swells up into rocky
ridges, or where from collapse of subterraneous cavities, in which
these regions abound, the strata are broken up and tumbled
about like heaps of ruins, there the rains sink too fast and so
deep that only the penetrating roots of the cedar can reach the
hidden moisture; a drought soon dries up the smaller herbage.
The cedars are always closely set, and it is a vexatious and ungrate-
ful task to penetrate such thickets.

In the oak barrens we find good farming lands as far as the sub-
carboniferous limestone extends. As soon as the siliceous or cherty

strata come to the surface—a good deal of such is in Tennessee—there we come into a poverty-stricken country. Like in the gravelly ridges of East Tennessee, so here too, the black-jack oak asserts its right. Spanish oak, sourwood and chestnut are the main body of the forest. Intervals are filled up with copses of sumach, dogwood, black haw, azaleas (*Azalea nudiflora* L.), *Kalmia latifolia* L., and various huckleberries (*Vaccinium arborum* Michx.), V. *stamineum* L., V. *corymbosum* L., Jersey-tea (*Ceanothus Americanus* L.). The herbaceous vegetation is monotonous, and in districts where the burning of the woods is practiced, of an unparalleled scarcity. Agricultural enterprise terminates very soon in the mutual ruin of land and farmer. Between such worthless lands are tracts or regions where the soil is of a yellow, light or fluffy loam, easily cultivated, not very rich, but apt to be kept in good condition. The subcarboniferous strata are here completely carried off and the Devonian strata become exposed and disintegrated into beds of loam or clay. We notice now a pleasant change in the appearance of well-kept farms, a better growth of timber and a much improved botanical prospect. Sometimes we pass by points where the subjacent strata are of an impervious clay, from which result heavy and damp soils, and in the early months of the year portions of the barrens are covered with shallow ponds, until they dry up in the hot season. Such spots are convenient abodes for orchids, liliaceous plants, Juncaceæ, Cyperaceæ, Gramineæ, Ludwigias, Rhexias, etc. The forest contains a good selection of hard-woods and the trees attain a stately growth. Water oak, willow oak and white oak, sweet gum and black gum are the most numerous. Ashes, poplars and beeches less frequent than in calcarious soils. The shrubbery is made up by alder (*Alnus serrulata* Ait.), willows (*Salix tristis* Ait., *S. humilis* Marsh.), botton-bush (*Cephalanthus occidentalis* L.), arrow-wood (*Viburnum nudum* L.), *Spiræa tomentosa* L., *Rosa Carolina* L., *Hypericum Kalmianum* L., *H. prolificum* L., *Comandra umbellata* Nutt.

I have appended for a ready review a comparative list of species of calcarious soils (glades) and siliceous soils (oak barrens).

PLANTS OF THE GLADES AND BLUFFS.

(Calcareous soils.)

Clematis reticulata Walt.
Thalictrum Cornuti L.
Anemone decapetala L.
Myosurus minimus L.
Ranunculus fascicularis Muhlb.
Delphinium azureum Michx.

Krigia Dandelion Nutt.
Prenanthes crepidinea Michx.
Lobelia Gattingeri Gray.
Bumelia lycioides Gært.
Forestiera acuminata Poir.
F. ligustrina Poir.

Magnolia acuminata L.
Calycocarpum Lyoni Nutt.
Cocculus Carolinus L.
Corydalis flavula DC.
Leavenworthia Michauxii Torr.
L. torulosa Gray.
L. stylosa Gray.
Draba brachycarpa Nutt.
Alyssum Lescurii Gray.
Cleome pungens Willd.
Lechea minor Walt.
Viola pubescens Ait.
V. striata Ait.
V. tricolor var. arvensis Gray.
Arenaria patula Michx.
Talinum teretifolium Pursh.
Hypericum aureum Bart.
H. sphaerocarpum Michx.
Malvastrum angustum Gray.
Sida Ellhottii Torr & Gray.
Ptelea trifoliata L.
Vitis indivisa Willd.
Aesculus glabra Willd.
Rhus aromatica Ait.
Trifolium reflexum L.
Psoralea subacaulis Torr. & Gray.
Petalostemon foliosus Nutt.
P. decumbens Nutt.
Desmodium pauciflorum DC.
D. Dilenii Darlingt.
D. Marylandicum Bart.
D. rigidum DC.
Baptisia australis R. B.
Gleditschia triacanthos L.
Desmanthus brachylobus Benth.
Prunus Chickasaw Michx.
Rosa humilis Marsh.
Crataegus cordata Ait.
Philadelphus hirsutus Nutt.
Sedum pulchellum Michx.
Oenothera triloba Nutt.
Melothria pendula L.
Trianosperma Boykinii Roem.
Ammania coccinea Rottboel.
Bupleurum rotundifolium L.
Eulophus Americanus Nutt.
Chaerophyllum Teinturieri Hook.
Houstonia patens Ell.
H. angustifolia Michx.
Galium virgatum Nutt.
Eupatorium incarnatum Walt.
E. altissimum L.
E. ageratoides L.
Grindelia lanceolata Nutt.
Solidago latifolia L.
S. Gattingerii Chapm.

Asclepiodora viridis Gray.
Asclepias variegata L.
As. verticillata L.
Acerates viridiflora Ell.
Sabbathia angularis Pursh.
Frasera Carolinensis Walt.
Phlox Stellaria Gray.
Nemophila microcalyx Fish & M.
Phacelia parviflora Pursh
Heliotropium tenellum Torr.
Lithospermum canescens Lehm.
Onosmodium Carolinianum DC.,
     var. molle Gray.
Evolvulus argenteus Pursh.
Cuscuta chlorocarpa Engelm.
Herpestes nigrescens Benth.
Gratiola Floridana Nutt.
Seymeria macrophylla Nutt.
Gerardia patula Chapm.
Catalpa speciosa Ward.
Ruellia ciliosa Pursh.
Isanthus coeruleus Michx.
Pycnanthemum linifolium Pursh.
Calamintha glabella Benth.
Monarda fistulosa L.
Aristolochia tomentosa Sims.
Oxybaphus albidus Sweet.
Euphorbia mercurialina Michx.
E. obtusata Pursh.
E. commutata Engelm.
Tragia macrocarpa Willd.
Croton capitatus Michx.
C. monanthogynos Michx.
Ulmus alata Michx.
U. racemosa Thom.
Quercus lyrata Walt.
Q. Mühlenbergii Engm.
Q. Michauxii Nutt.
Q. imbricaria Michx.
Orchis spectabilis L.
Spiranthes gracilis Bigelow.
Pardanthus Chinensis Kerr.
Allium mutabile Michx.
Schoenolirium croceum Gray.
Uvularia grandiflora Smith.
Juncus leptocaulis Torr. & Gray.
Cyperus Baldwinii Torr
C. Engelmanni Hend.
C. acuminatus Torr.
C. ovularis Torr.
Eleocharis compressa Sulln.
Scirpus lineatus Michx.
Carex Muhlenbergii Schk.
C. retroflexa Mühll.
C. Crawei Dew.
Paspalum dilatatum Poir.

S. ulmifolia Nutt.
S. rupestris Raf.
Bellis integrifolia Michx.
Aster oblongifolius Nutt.
A. Shortii Hooker.
A. undulatus L.
A. lævis L.
Polymnia Canadensis, var. radiata Gray.
Silphium perfoliatum L.
S. terebinthinaceum, var. pinnatifidum Gray.
Inula Helenium L. E. Ten.
Echinacea angustifolia DC.
Lepachys pinnata Torr & Gray.
Helianthus hirsutus Raf.
H. tuberosus L.
Verbesina Virginica L.
Cacalia tuberosa Nutt.

Panicum latifolium L.
P. capillare L., var. flexile, Gattinger.
P. dichotomum L., var. nitidum Lam..
Tripsacum dactyloides L.
Danthonia sericea Nutt.
Eatonia Dudleyi Vasey.
E. Pennsylvanica Gray.
Eragrostis Frankii Meyer.
Aristida gracilis Ell.
Sporobolus vaginæflorus Torr.
Uniola latifolia Michx.
Festuca ovina L.
Arundinaria macrosperma Mich.
Juniperus Virginiana L.
Cheilanthes Alabamensis Kunze.
Ch. vestita Swartz.
Asplenium parvulum M. & J.
Isœtes Buttleri, var. immaculata Engelm.

PLANTS OF THE OAK BARRENS AND HIGHLANDS.

(Siliceous and Argillaceous soils.

Thalictrum dioicum L.
Th. debile Buckl.
Ranunculus oblongifolius Ell.
Delphinium tricorne Michx.
Cimicifuga racemosa Ell.
Caulophyllum thalictroides Michx.
Stylophorum diphyllum Nutt.
Dicentra Cucullaria DC.
Vesicaria Shortii Torr & Gray. (on shale.)
Polanysia graveolens Raf.
Helianthemum Canadense Michx.
Lechea patula Legget.
L. tenuifolia Michx.
Viola pedata L.
V. blanda L.
Polygala Curtissii Gray (and other Polygalas).
Anychia dichotoma Willd.
Hypericum virgatum Lam.
H. Kalmianum L.
H. dolabriforme Vent., (E. Ten.).
H. nudicaule Walt.
Vitis vulpina L.
Rhus copallina L.
Crotallaria sagittalis L
Psoralea melilothoides Michx.
Stylosanthes elatior Swartz.
Lespedeza striata Hook.
L. capitata Michx.
L. hirta Michx.
Thephrosia Virginica Pers.
Th. spicata Torr. & Gray.

Krigia Virginica Pursh.
Lobelia puberula Michx.
Vaccinium arboreum Marsh.
V. stamineum L.
Oxydendron arboreum DC.
Kalmia latifolia L.
Rhododendron nudiflorum Torr.
Rh. calendulaceum Torr.
Acerates longifolia Ell.
Sabbathia gracilis Pursh.
Phlox amœna Sims.
Lithospermum latifolium Michx.
Convolvulus spithameus L.
Cuscuta compacta Iuss.
C. glomerata Choisy.
Gratiola ramosa Walt.
Buchnera Americana L.
Seymeria tenuifolia Pursh., E. Ten.
Gerardia quercifolia Pursh.
G. tenuifolia Vahl.
Castilleia coccinea Spreng.
Schwalbea Americana Gron.
Trichostema dichotomum L.
Pycnanthemum lanceolatum Pursh.
P. Tullia Benth.
Monarda Bradburiana Buckl.
Oxybaphus nyctagvineus Sweet.
Comandra umbellata Nutt.
Pachysandra procumbens Nutt.
Euphorbia corollata L.
Quercus falcata Michx.
Q. aquatica Cate-by.
Q. Phellos L.

Rhynchosia tomentosa Torr.& Gray.
Desmodium viridiflorum Buckl.
D. sessilifolium Torr.
Clitoria Mariana L.
Baptisia tinctoria R. Br.
Thermopsis Caroliniana M. A. Curt.
Cladrastis tinctoria Raf.
Schrankia angustata Torr. & Gray.
Œnothera fruticosa L.
Œ. sinuata L.
Ludwigia hirtella Raf.
L. linearis Walt.
Hamamelis Virginica L.
Lythrum alatum Pursh.
Gaura biennis L., (E. Ten.).
G. filipes Spach, (E. Ten.).
Polytæia Nuttallii DC.
Archangelica hirsuta Torr. & Gray.
Houstonia cærulea L.
Oldenlandia Boscii Chapm.
Eupatorium semiserratum DC.
E. aromaticum L.
E. leucolepis Torr. & Gray.
Liatris spicata Willd.
L. graminifolia Pursh.
Chrysopsis Mariana Nutt.
Solidago cæsia L.
S. bicolor L.
S. odora Ait.
S. corymbosa Ell.
S. lanceolata L.
Boltonia diffusa Ell.
Aster concolor L.
A. umbellatus Mill.
A. linariifolius L.
Silphium brachiatum Gattinger.
S. scaberrimum Ell.
S. Asteriscus Ell.
Rudbekia laciniata L.
Helianthus atrorubens L.
Helianthella tenuifolia Torr.& Gray.
Coreopsis auriculata L.
C. senifolia Michx.
Cacalia suaveolens L.

Castanea vesca, var. Americana
    Michx.
C. pumila Michx. (E. Tenn.).
Habenaria integra Sprengel.
H. peramœna Gray.
Spiranthes simplex Gray.
Sp. cernua Rich.
Pogonia pendula Lindley.
Iris Virginica L
Lilium Canadense L.
Uvularia perfoliata L.
Stenanthium robustum Wats.
Amianthium muscœtoxicum, Gray,
    (E. Tenn.).
Dichromena latifolia Baldso.
Juncus pelocarpus E. Meyer.
Cyperus vegetus Willd.
Scirpus polyphyllus Vahl.
Carex Shortiana Dev.
C. flaccosperma Dew.
C. granularis Mühlb.
Paspalum læve Michx.
Panicum commutatum Schulz.
P. depauperatum Mhlb.
P. filiforme L.
P. glabrum Gaud., var. Mississippi
    ense Gattinger.
Erianthus alopecuroides Ell.
E. brevibarbis Michx.
E. strictus Baldw.
Danthonia spicata Beauv.
Gymnopogon racmeosus Beauv.
Eatonia obtusata, var. laxiflora Gatt.
Chryspogon avenaceum Buckl.
Aristida purpurascens Poir.
Sporobolus Indicus R. Br.
Uniola gracilis Michx.
Festuca Shortii Vasey.
Arundinaria tecta Mhlb.
Pinus inops Ait.
Equisetum arvense L., E. Tenn.
Pteris aquilina L.
Phegopteris hexagonoptera Feè.

WEST TENNESSEE.

The Tennessee river very nearly indicates in its northern course a geological division, flowing, as it does, along an ancient devonian and silurian shore-line. A few miles west and parallel with the river rises the eastern escarpment of an undulating plateau of only 2–300 feet elevation above the waters of the Tennessee river. This irregular tableland slopes gradually towards the Mississippi river and terminates there in another bluff, which rises about 200 feet over the floods of the Mississippi. The eastern portion of this area is compo ed of cretaceous and the western of tertiery and post-tertiery deposits, either sands or soft cretaceous shale. Solid, often ferruginous, sandstones appear at the surface, scattered in incoherent masses.

We behold no longer limpid streams, rippling over rocky bottoms, sided by cliffs and bluffs. Instead of them we find lagoons and swampy borders, stretching along muddy-looking waters of sluggish streams.

From distance already, before crossing the river, we are in sight of towering cypresses. While a thousand miles east from here they yet occupy the shore-line of the Atlantic, here the shore line has receded to the Gulf and left the cypress behind. Their dimensions are truly enormous. The far spreading roots emerge like sharp-backed ridges from the brownish lagoon, gradually creeping up and girding with buttress-like projections the many-angled column. A perpendicular shaft ascends to a height of 120–150 feet and then spreads a flat or hemispherical crown. Such I have seen twenty years ago near Johnsonville. Cypress swamps are along both big rivers, and many other extensive swamps and swampy lands are along every watercourse, the most perhaps along Big Sandy. It may therefore be expected that a great many more aquatic species and such as inhabit marshy lands exist in this region, than in either East or Middle Tennessee. My own experience is, however, limited and restricted to one point on the Mississippi river, the regions of Brownsville, Humbolt, McKenzie, Hollow-rock and Johnsonville, in which places I have made interesting collections.

PLANTS PECULIAR TO WEST TENNESSEE.

Ranunculus multifidus Pursh.
Brasenia peltata Pursh.
Cabomba Caroliniana Gray.
Hypericum lobocarpum Gattinger.
Gleditschia monosperma Walt.
Dalea alopecuroides Willd.
Berchemia volubilis DC.
Galium Arkansanum Gray.
Eryngium prostratum Nutt.
Marshallia lanceolata Pursh.
Ambrosia bidentata Michx.
Helenium tennifolium Nutt.
Senecio lobatus Pers.
Hydrolea affinis Gray.
Verbena stricta Vent.
Limnanthemum lacunosum Griese.
Polypremum procumbens L.
Quercus bicolor Willd.
Stillingia sylvatica L.

Planera aquatica Gmel.
Lithospermum anguistifolium Michx.
Utricularia biflora Lam.
Iris cuprea Pursh.
Iris hexagona Walt.
Habenaria virescens Sprengel.
Juncus mititaris Bigel.
Scirpus debilis Pursh.
Zizania aquatica L.
Spartina cynosuroides Willd.
Eragrostis oxylepis Torr.
Cenchrus tribuloides L.
Aristida ramosissima Engelm.
Taxodium distichum Rich.
Asplenium Filix femina L. var. angustum Gray.
Azolla Caroliniana Willd.
Equisetum robustum R. Br.

Tennessee Flora.

Species occurring within a radius of thirty miles from
Nashville are reckoned to the Nashville Flora,
and printed in bold type; those beyond
this range in small pica. O. S.—
Over the whole State.

RANUNCULACEÆ.

Clematis Viorna L. O. S. Very variable.
Var. *coccinea* Gray. Foot of Lookout Mt. J. F. James.

C. reticulata Walt. Middle Tenn. Foliage not quite as
leathery and prominently reticulated as my S. C. specimens.

C. Virginiania L. O. S.

Anemone nemorosa L. Paradise ridge. Cumberland and
Alleghany Mts.

A. Virgianiana L. O. S. June–July.

A. acutiloba Lawson. Middle Tenn. March.

A. Hepatica L. Mts. of East Tenn. In the Frog Mts.,
East Tenn., I obtained specimens with acute and obtuse-lobed
foliage on the same plant. April.

A. decapetala L. (A. Caroliniana Walt.) Cedar glades. April.

Anemonella thalictroides Spach. (Th. anemonoides Michx.)
O. S. March.

Thalictrum dioicum L. O. S. June–July.

T. purpurascens L. O. S. July.
Var. **ceriferum** Aust. O. S. June–July.

T. clavatum DC. Cumberland and Alleghany Mts. July.

T. polygamum Mühlb. Moist woodlands. O. S. July–August.

T. debile Buckl. Dr. Hampton's place near Cheatham Co. line. Only locality. May.

Trautvetteria palmata Fish & Meyer. Along the whole chain of Alleghanies. July.

Myosurus minimus L. Moist meadow-lands. Nashville. April.

Ranunculus abortious L. *var. micranthus* Gray; common. March.

R. ambigens Watson. Ponds along Cumberland river, etc. June–July.

R. circinatus Sibth. (R. aquatilis L. var. stagnatilis DC.) Ponds and swamps along Cumberland R. Flowers the whole summer.

R. aquatilis L. var. trichophyllus Chaix. East Tenn. July.

R. fascicularis Mühlb. Common in vicinity of Nashville. March–April.

R. multifidus Pursh. Swamps on Cumberland and Tenn. rivers. Cypress swamps. July.

R. oblongifolius Elliott. Damp ground. Paradise ridge. Tullahoma. June.

R. parviflorus L. Boggy lands. O. S. April–May.

R. pusillus Poir. Ditches and waste grounds. May.

R. septentrionalis Poir. var. hispidus, Michx. (R. repens L. var. hispidus Michx.) O. S.
Var. lucidus Poir. O. S. May–July.

Caltha palustris L. var. *parnassifolia* Torr & Gray. Wet mountain meadows, etc. Ducktown. April.

Aquilegia Canadensis L. Rocky woodlands. Bluffs on Mill Creek near Nashville. April.

Delphinium azureum Michx. Cedar glades of Middle Tenn. May.

D. tricorne Michx. Rich woodlands. O. S. April–May.

D. exaltatum Ait. Roane Mt. J. W. Chickering.

D. Consolida L. Introduced and spreading copiously. May, June.

Aconitum reclinatum Gray. Mountains of East Tenn.

Coptis trifolia Salisb. Higher Alleghanies. Thunderhead. June.

Xanthorrhiza apiifolia L'Her. Banks of every mountain stream. East Tenn. March–April.

Hydrastis Canadensis L. Rich woods; very common. April, May.

Actaea alba Bigel. Rich woodlands. May.

A. spicata var. rubra Michx. Moist, rich and shady localities. Sometimes with the former. O. S.

Cimicifuga racemosa Ell. Rich woodlands; common. June–July.

C. Americana Michx. Alleghany Mts., Ducktown, Roane Mt. Prof. Chickering. August–September.

CALYCANTHACEÆ.

Calycanthus floridus L. Harpeth hills by Nashville. Frequently cultivated in old gardens. May.

C. glaucus Willd. Cumberland Mts. Whiteside. Also Chilhouewe and Smoky Mts. June.

MAGNOLIACEÆ.

Magnolia acuminata L. River and creek bottoms of Middle Tenn. Along Gallatin pike, near Nashville. May.

M. macrophylla Michx. Mountains of East Tenn., Smoky Mts., etc. May–June.

M. Umbrella L. Cumberland Mts., in deep ravines: also in Blue ridge. June.

M. Fraseri Walt. Apparently not in the Cumberland Mts., but frequent in high mountains of East Tenn., Roane Mts., Big Smokies. June–July.

M. grandiflora L. In cultivation only. May–July.

Liriodendron tulipifera L. One of the tallest trees in the State. The largest trees are thought to be found along Mississippi bottoms in Obion Co., and to measure above 150 feet high. May.

ANONACEÆ.

Asimina triloba Dunal. Commonly a shrub 8–10 feet high, but also found a small, slim tree 6–8 inches diameter and 25 feet high, in the rich bottoms of Cumberl. and Tenn. rivers. March.

MENISPERMACE.E.

Calycocarpum Lyoni Nutt. River bottoms, climbing high. May–June.

Cocculus Carolinus L. On bushes in open fields and climbing high in moist woodlands. July–August.

Menispermum Canadense L. River bottoms, with the former. July.

BERBERIDACE.E.

Caulophyllum thalictroides Michx. In deep leaf-mould. Frequent in Cumberland Mts and Alleghanies. April.

Diphylleia cymosa Michx. Roane Mts. Prof. Chickering. Also Smoky Mts. May–June.

Jeffersonia diphylla Persoon. Rocky woodlands, over the State. In limestone soil. Harpeth ridge. May.

Podophyllum peltatum L. Rich woodlands, every where. April–May.

NYMPHÆACE.E.

Brasenia peltata Pursh. Ponds and lagoons. Not frequent. Swamp in Jones' Bend, near Edgefield Junction. July.

Cabomba Caroliniana Gray. Cypress swamps, Johnsonville, W. Tenn. July–August.

Nelumbium luteum Willd. Lagoons along Tennessee and Cumberland rivers. Shelby Pond, near Nashville. July.

Nuphar advena Ait. Very frequent in the lower course of Cumberland Mts. streams. June–Sept.

Nymphæa odorata Ait. In a pond at the Lunatic Asylum grounds near Nashville. Said to be in the State.

PAPAVERACE.E.

Argemone Mexicana L. O. S., in and near towns and villages.

Stylophorum diphyllum Nutt. Rich woods. Harpeth hills, near Nashville. April–May.

Papaver somniferum L. Occasionally escaped, but not inclined to spread. July–August.

P. dubium L. Waste grounds in Nashville. Old cemetery. June.

Sanguinaria Canadensis L. Over the whole State in rich soil. April–May.

FUMARIACEÆ.

Dicentra Cucullaria DC. In leaf mould, shady ravines. Harpeth hills near Nashville. April.

D. Canadensis DC. Cumberland mountains. Grab Orchard, etc. May.

D. eximia DC. Only on Dow river, Carter Co., but there abundant. June–July.

Adlumia cirrhosa Raf. In the valley of Dow river, along narrow gauge railroad, leading to Craneberry iron works, abundant.

Corydalis flavula DC. Woods and thickets. Nashville.

C. glauca Pursh. Mountain gorges on Dow river. East Tenn. August.

CRUCIFERÆ.

Nasturtium officinale R. Br. Springs and brooklets. East and Middle Tenn. April–May. Indigenous.

N. lacustre Gray. Swamps of Tennessee and Cumberland rivers. (Johnsonville). July.

N. sessiliflorum R. Br. Wet meadows and ditches. Common. May–August.

N. Armoracia Fries. Horseradish. In cultivation, and here and there near gardens.

Leavenworthia Michauxii Torr. Cedar glades. Middle Tenn. Lavergne. April–May.

*L. torulosa Gray. n. sp. First collected in 1865 in vicinity of Vanderbilt University grounds. Abounds about Nashville and over the cedar glades of Middle Tenn. April.

---

*Leavenworthia torulosa Gray. Silique linear, conspicuously torose; style fully equalling the breadth of the silique; seeds broadly oval, narrowly winged ; radicle nearly transverse, strictly applied to the edges of the cotyledons at the base on one side; petals purplish with a yellowish spot towards the claw. Either stemless or caulescent, ascending from a spreading procumbent base, with several sometimes ultimately forking pedicels. Generally 4-5, but large specimens sometimes 9 inches high, and spreading over a square foot of ground. The fresh herbage has the taste of water-cress and is well adapted for table use. Vide *Bot. Gaz.*, March, 1880.

*L. stylosa Gray.   n. sp.   Discovered in the cedar glades 1 mile east of Lavergne, 17 miles from Nashville, in 1869.   Also found near Green Hill, in Wilson Co., Tenn.   April.

Dentaria diphylla Michx.   Banks of Cumberland, Nashville. April.

D. laciniata, Mühlb.   Hills around Nashville.   March–April.

D. multifida, Mühlb.   East Tenn.   Roane Co.   April.

Cardamine rhomboidea DC.   Low grounds.   Vicinity of race-track at Nashville.   May.

C. Clematitis Shuttl.   Highest mountains of East Tenn. Roane Mt.   Prof. Chickering.   Clingman Dom 6500'.   June–July.

C. hirsuta L.   Low swampy grounds.   April–May.
   Var. sylvatica Gray.   Dry woodlands.   March–May.

Arabis Ludoviciana Meyer.   Fields and roadsides, abundant. March, May.

A. dentata Torr & Gray.   Low, rich grounds.   O. S.   March– May.

A. Canadensis Liv.   Rocky woodlands.   O. S.   April.

A. patens Sulliv.   Along Ocoë river, East Tenn., etc.

A. lævigata DC.   Rocky woodland, cliffs along Cumberland and Mill creek.   April–May.

Thelipodium pinnatifidum Wats   Rich hillsides.   Frequent around Nashville.   April–May.

Brassica Sinapistrum Boiss.   In cultivated grounds.

---

*Leavenworthia stylosa Gray.   Habitus the same like the former, but a little more slender.   It is also either stemless or caulescent, not strictly stemless, like Dr. Gray thought it to be from stemless specimens sent to him.   Robust plants have the ascending stems terminating with a fasciculate inflorescence.   Silicle broadly oval or oblong and 2 lines wide and 5–12 lines long, plane, surmounted by a slender style 3–4½ lines long : Seeds 3–6 orbicular, distinctly winged ; embryo as in the preceding : petals during anthesis pure golden yellow throughout, marcescent and shrivelling they turn to a purplish-white.

First collected in the cedar barrens at Lavergne, one mile southeast from the station, June 2, 1879, in a springy spot associated with Isoetes Buttleri and Schœnolirium Carolinianum.   Also found in a similar spot near Green Hill, Wilson Co.

Leavenworthia Michauxii Torr. occurs in the same locality close by, is always much smaller ; stigma sessile or subsessile.   The name Cardamine uniflora may have originated in an oversight of robuster specimens or a too early collection.   Vide *Bot. Gaz.*, March, 1880.

B. alba Gray. Frequently cultivated and escaping. May.

B. nigra Koch. Escaped.

B. campestris. Cultivated and escaping into waste grounds.

Draba brachycarpa Nutt. Cedar glades of Middle Tenn. March.

D. Caroliniana Walt. Sterile rocky lands. O. S. March–April.

Draba ramosissima Desv. Mountains of East Tenn. Mundic bluff on Ocoë river, Polk Co. April–May.

Vesicaria Shortii Torr & Gray. Rising Sun bluff, 14 miles below Nashville. April.

Alyssum Lescurii Gray. Hills south and west of Nashville, covering the ground in great patches, visible miles distant. April–May.

Sisymbrium officinale Scop. Ditches and roadsides, everywhere. May–June.

Thlaspi arvense L. Common weed. March–June.

Hesperis matronalis L. Introduced. Alongside a fence in Mrs. Cheatham's gardens, Nashville. May.

Lepidium Virginicum L. O. S., along roadsides. May–July.

Senebiera Coronopus L. Vacant town lots, Memphis. Dr. G. Egeling.

Raphanus sativus L. Escapes sometimes into open fields.

CAPPARIDACEÆ.

Cleome pungens Willd. Scattered O. S. Abundant in low woodlands along Tennessee river near Chattanooga. June–July.

Polanisia graveolens Raf. Along R. R. in Dixon Co., Middle, Tenn. Also West Tenn. July.

CISTACEÆ.

Helianthemum Canadense Michx. Bradley Co., East Tenn. near Charleston. April–May.

Lechea major Michx. O. S. July–August.

L. minor Walter. Barrens of Middle Tenn. and mountains of East Tenn. July–August.

L. tenuifolia Michx. Cumberland and Alleghany Mts. July–August.

L. thymifolia Pursh. Dry rocky glades, Middle Tenn. July–August.

L. Drummondii Torr. & Gray. Cedar glades, Lavergne. May–June.

VIOLACEÆ.

Ionidium concolor Barth. & Hook. O. S. April–May.

Viola pubescens Ait. Harpeth hills, near Nashville, etc.
Var. **eriocarpa** Nutt. Rich woods of the Cumberland Mts. and banks of Cumberland river at Nashville. April.

V. canina L. var. **sylvestris** Regel. Damp woodlands, East Tenn. March–April.

V. striata Ait. O. S. Frequent in vicinity of Nashville. April–May.

V. Canadensis L. Cumberland and Alleghany Mts. May–July.

V. pedata L. Over the whole State.
Var. **bicolor** Pursh., with the former. April–May.

V. sagittata Ait. Highlands of Middle Tenn. East Tenn. April–May.

V. blanda Willd. Dry rocky ground, hills near Nashville. April–May.
Var. *renifolia* Gray. Summit of Thunderhead, 6000′. July.

V. rotundifolia Michx. Cumberland Mts.

V. palmata L. Common. April–May.
Var. **cucullata** Ait. With the former. April–May.

V. odorata L. Old cemeteries and escaped from gardens. Not prone to spread.

V. tricolor L. var. **arvensis** Gray. Abundant in the cedar glades of Middle Tenn. March–April. Indigenous.

POLYGALACEÆ.

Polygala incarnata L. Paradise ridge, Tullahoma, Bon Air. June–July.

P. Curtissii Gray. Barrens of Middle Tenn. to Mts. of East Tenn. June–July.

Polygala fastigiata Nutt. Sewanee, Tenn. July–August.

P. cruciata L. Oak barrens. O. S. July–August.

P. Nuttallii Torr. & Gray. Argillaceous and siliceous soils. O. S. July–August.

P. ambigua Nutt. Dry, sterile and siliceous soils. Harpeth hills and vicinity of Nashville to East Tenn. July.

P. polygama Walt. East Tenn. Frequent.

P. Senega L. East Tenn.

Var. **latifolia** Torr. & Gray. Vicinity of Nashville. May–June.

CARYOPHYLLACEÆ.

Saponaria officinalis L. Introduced but now divulged over the whole State. July–August.

Silene antirrhina L. Fields and waste places. May.

S. Pennsylvanica Michx. Cumberland and Allegheny Mts. May.

S. Virginica L. Edges of woods and sunny hillsides. May.

S. rotundifolia Nutt. High cliffs of Cumberland Mts., on sandstone ledges near Sewanee. Lookout Mt., summit of. May–June.

S. stellata Ait. Open woodlands. Common. July.

S. nivea DC. Mountains on Emory river above Kingston, East Tenn. Apparently very rare!

Lychnis Githago Lam. In wheat and rye-fields. May.

Arenaria glabra Michx. Highest parts of Cumberland and Allegheny Mts. July. (Lookout Mt.)

A. patula Michx. Cedar Glades of Middle Tenn., also Knoxville, East Tenn. April–May.

A. serpyllifolia L. Dry, rocky grounds. March–May.

A. diffusa Elliott. Rich, shady soil. May–June.

Stellaria media Smith. Troublesome weed in cultivated grounds. Flowering summer and winter.

S. pubera Michx. Rich woodlands. April–May.

S. crassifolia Ehrh. Moist and rocky grounds. Very abundant in vicinity of Nashville. April–May.

S. longifolia Mühlb. Kingston springs on the bluff. July.

Cerastium vulgatum L. Pastures and roadsides. April–May.

C. viscosum L. Less copious than the former. Dry uplands. April–May.

C. oblongifolium Torr. Cedar glades. May–June.

C. nutans Raf. Moist ground and hillsides. May–June.

Sagina apetala L. Damp soil. Pavements in Nashville. March–April.

PARONYCHIEÆ.

Paronychia argyrocoma Nutt. High mountains of East Tenn. August. Prof. Chickering.

Anychia dichotoma Willd. Dry rocky woodlands.

A. capillacea DC. With the former. July–August.

PORTULACACEÆ.

Portulaca oleracea L. Fields and gardens. June–September.

P. grandiflora Hook. Frequently cultivated in flower-gardens and therefrom self-sowing. June–September.

Talinum teretifolium Pursh. Crevices of rocks in the cedar glades. July–August.

Claytonia Virginica L. Woods and pastures. March–May.

C. Caroliniana Michx. Mountains of East Tenn. Ducktown. May.

HYPERICINEÆ.

Ascyrum Crux Andreæ L. Siliceous formation. July–Sept.

A. stans Michx. Mountain bogs. July–August.

A. hypericoides L. Cleveland, East Tenn., Chilhowee Mts. June–July.

Hypericum prolificum L. East Tenn., Middle Tenn. and Craggy-hope, Dixon Co. June–July.

Var. *montanum* Mihi. Frog Mts., Polk Co. July.

*H. lobocarpum, n. sp. Hollow Rock, West Tenn. July.

---

* H. *lobocarpum* Gattinger, n. sp. Sepals linear-lanceolate, small, unequal, 1½–3 lines long. Petals unequal, unsymmetric, 3–6 lines long, reflected, early deciduous. Capsule five-celled, deeply five-lobed, lanceolate, tapering into a long beak. Carpels almost distinct, and at full maturity falling away from a central axis. Seeds 1 mm. long, incurved, apiculate, striate lengthwise, transversely groove t. Leaves linear, obtuse, slightly mucronate, attenuate downwards, pale underneath.

Shrub, 5–7 feet high, with upright branches. Low swampy lands in the orange sand formation at Hollow Rock, Carroll Co., West Tennessee. First collected in fruit in 1867, and again July, 1886, in flower. Only two shrubs found, in very swampy ground, at the time nearly inaccessible. I have since received specimens of a Hypericum labelled H. prolificum, "collected by Dr. H. E. Hasse of Little Rock, Ark., in wet pine barrens," which prove to be the same species.

H. Kalminanum L. var. majus. Barrens at Tullahoma. July. 5–6 feet high and of more robust habit than the northern forms.

H. aureum Bartram. Limestone ledges, Mts. at Cowan ; frequent in the cedar glades and upon cliffs on Cumberland river. July.

H. fasciculatum Lam. In springy places, forming thickets. Bradley Co., East Tenn. July.

H. adpressum Barton. Wild goose pond near Mitchelville, Summer Co. August–September.

H. sphaerocarpum Michx. Moist places in the barrens. Frequent around Nashville. June–July.

H. dolabriforme Vent. Cameron Hill, Chattanooga. July.

H. virgatum Lam. (*H. angulosum* Michx.) Barrens of Middle Tenn., damp argillaceous soils. July–September.

Var. acutifolium. With the preceding.

H. perforatum L. Waste grounds ; not frequent here. June–July.

H. corymbosum Mühl. O. S. July.

H. graveolens Buckl. Summit of Clingman Don, Smoky Mts., East Tenn. 6000'. July.

H. mutilum L. Very abundant in wet places. July–Sept.

H. gymnanthum Engelm. & Gray. Barrens at Tullahoma. July.

H. Canadense L. Cumberland Mts. July–August.

H. Drummondii Torr. & Gray. Belvidere, Franklin Co. July–August.

H. nudicaule Walter (*H. Sarothra* Michx.). In sandy soil everywhere. July.

Elodea campanulata Pursh. *E. Virginica* Nutt. River—and swamps near Nashville. July–August.

E. petiolata Pursh. Swamps, Middle and West Tenn. July–September.

TERNSTRŒMIACEÆ.

Stuartia pentagyna L'Her. Cumberland and Allegheny Mts. June.

MALVACEÆ.

Abutilon Avicennæ Gaertn.   Waste places.   July–August.

Malva rotundifolia L.   Around dwellings.   June–September.

Callirrhœ alcæoides Gray.   Copses near Brown's creek.   Nashville.   Found only once.

Napaea dioica L.   Upper East Tenn. near Johnson City.   June–August.

Malvastrum angustum Gray.   Rocky ground and open glades.   Middle Tenn.   Very abundant.   July–August.

Sida Elliottii Torr. & Gray.   Cedar glades.   Lavergne.   Edgefield Junction.   July–August.

S. spinosa L.   Troublesome weed in pastures and fields.   July–September.

Hibiscus Moshentos L.   River banks and swamps.   July.

H. militaris Cav.   River banks and bottom lands.   A white variety near Nashville.   June–July.

H. Trionum L.   Frequently cultivated in flower gardens, and thence escaped.   July–August.

TILIACEÆ.

Tilia Americana L., var *pubescens* Land.   Mountains of East Tenn.   Ocoë river.

T. heterophylla Vent.   Rich woodlands and along river banks.   Nashville.

LINACEÆ.

Linum Virginianum L.   Dry open woodlands.   June.

L. striatum Walt.   Cedar and oak barrens.   June.

L. usitatissimum Linn.   Waifs occasionally found near dwellings.

L. sulcatum Riddel.   Near Hickman, and East Tenn.   July.

GERANIACEÆ.

Geranium maculatum L.   Rich woodlands.   June–July.

G. Carolinianum L.   Waste places.   Common.   May–June.

Floerkea proserpinacoides Willd.   Low moist grounds.   Valley of East Tenn.   April–June.

Impatiens pallida Nutt. Thickets along creeks and springs. July–September.

I. fulva Nutt. With the former. O. S. June–September.

Oxalis Acetosella L. Summit of high Mts. of East Tenn. July. (Thunderhead, Roane Mt.)

0. violacea L. Rocky places. Very common. May–June.

0. stricta L. Copses and margins of woodlands. Very variable in size of flowers. May–September.

RUTACEÆ.

Xanthoxylum Americanum Mill. Hills vicinity of Nashville. (Charlotte pike, copses beyond Edgefield, etc.) April–May.

Ptelea trifoliata L. A common shrub in the limestone regions of Tennessee. May–June.

SIMARUBEÆ.

Ailanthus glandulosa Desf. Perfectly naturalized. Widely spreading over the State. May.

MELIACEÆ.

Melia Azedarach L. Formerly frequently planted around dwellings, and sometimes found on deserted homesteads. It is gradually dying out. Better adapted is

Kœhlreutera paniculata DC. Lately introduced.

ILICINEÆ.

Ilex opaca Ait. Mountains. Cumberland and Alleghenies. Also in islands of Cumberland and Tennessee rivers, where it grows to a stately tree with 20″ diameetr and 40–50 feet high. June.

Ilex decidua Walter. Brownsville, West Tenn. May.

I. mollis Gray. Cumberland Mts. Summit of Lookout Mt. May.

I. monticola Gray. Smoky Mts., East Tenn. May.

Nemopanthes Canadensis DC. Mts. at Cowan. August.

CELASTRACEÆ.

Celastrus scandens Linn. Brownsville, West Tenn. May–June.

Evonymus Americanus L. Rich woodlands. June.

E. atropurpureus Jacq. River banks and copses. June–July.

RHAMNACEÆ.

Berchemia volubilis DC.   Lookout Mt.   Very luxuriant and frequent in West Tenn.   Brownsville.   May–June.

Rhamnus lanceolata Pursh.   River banks and copses, Nashville.

Rh. Caroliniana Walt.   Barrens of Middle Tenn, etc.   June.

Ceanothus Americanus L.   In argillaceous and siliceous soils, woodlands throughout the State.   June–July.

VITACEÆ.

Vitis Labrusca L.   Alleghany Mts., Ducktown.   Cultivated in Bayers settlement by the first settlers.   June.

V. æstivalis Michx.   Uplands, cedar glades.   Also in Alleghany Mts.   May.

V. cordifolia Michx.   Along water courses.   April–May.

V. riparia Michx.   With the former.   May.

V. vulpina L.   Cumberland and Alleghany Mts., abundant. April–May.

V. rupestris Trelease.   Along Cumberland river, banks of Stoner's creek etc., Nashville.   May.

V. arborea Linn.   West Tenn. in moist woodlands.   May–June.

V. indivisa Willd.   River banks and woodlands, Middle Tenn. May.

Ampelopsis quinquefolia Michx.   Common.

SAPINDACEÆ.

Acer dasycarpum Ehrh.   Low, damp grounds and bottom lands.   Frequently planted for shade-trees.   Flowers in February and matures seeds before any other plant.

A. rubrum L.   Wet and low lands.   Also cultivated for shade-trees.   April.

A. saccharinum Wanger. var. **nigrum** Torr. & Gray.   Rich soils.   O. S.

A. spicatum Lam.   Smoky Mts.   Summit of Thunderhead. June.   A middle-sized tree.

Negundo aceroides Mœnch. Banks of creeks and rivers, growing to large dimensions but being unshapely. April.

Æsculus glabra Willd. Very abundant in the barrens of Middle Tenn. May.

Æ. flava Ait. Rich woods, especially in the mountains. April.

Var. **purpurascens,** small shrub, loving ravines and shady gorges. East Tenn. May.

*__Æ. Pavia__ L. Prospect Station, Giles Co., Tenn.

Cardiospermum Halicacabum L. Waste places around dwellings. June–September.

Staphylea trifolia L. Thickets in rich, moist soil. April.

ANACARDIACEÆ.

Rhus typhina L. Hillsides. June.

R. glabra L. Poor soils and deserted farmlands. June.

R. copallina L. Hills and rocky siliceous soils. July.

R. venenata DC. Swampy or boggy lands, especially in the mountains. June.

R. Toxicodendrum L. In all woodlands. June.

R. aromatica Ait. Limestone regions of Middle Tenn. April–May.

LEGUMINOSÆ.

Lupinus perennis L. Rich hillsides, Ducktown, East Tenn. (Smelting works at Hiwassee.) May.

Crotolaria sagittalis L. Sandy soil, Tullahoma, Paradise ridge. June–July.

Trifolium arvense L. Old fields and pastures. June–July.

T. pratense L. Largely cultivated and frequently spontaneous. July.

T. reflexum L. Cedar barrens, rocky hills along Franklin pike, Nashville. April–May.

T. repens L. Common everywhere. May.

T. procumbens L. Very abundant over the whole State, in old fields and pastures. May.

---

*Æ-culus Hippocastanum L. Is sometimes planted, but entirely unadapted to our climate.

3

T. stoloniferum Michx. Found in one locality where reflexum abounds (Franklin pike). It is perhaps only a form of reflexum or a hybrid. May–June.

Melilothus alba Lam. Around dwellings, vacant town lots and grass plots in Nashville. June–July.

M. officinalis Willd. Waste grounds. May–June.

Medicago lupulina L. Grass plots in Nashville, Capitol grounds. June.

M. sativa L. Occasionally cultivated, becoming naturalized. July.

Psoralea Onobrychis Nutt. Paradise ridge, etc. Not frequent. June.

P. melilotoides Michx. Hills around Nashville. May–June.

P. subacaulis Torr. & Gray. Cedar glades and rocky hills around Nashville. April–May.

Dalea alopecuroides Willd. Frequent in West Tenn. July–August.

Petalostemon violaceus Michx. Davidson's farm, in a cedar glade on the Charlotte pike, near Nashville. June.

P. candidus Michx. Near tunnel at Cowan; also Paradise ridge near Nashville. July.

P. foliosus Nutt. Frequent in the cedar glades around Nashville. July.

P. decumbens Nutt. Lavergne, Trabue's place on Mill creek, near Nashville, and other points in the cedar barrens. A white variety occurs at Lavergne. June.

Amorpha fruticosa L. Banks of Cumberland river, Nashville, etc. May.

A. Tennessiensis Shuttlw. High mountains of East Tenn. and on the banks of mountain streams. Parksville, Polk Co.; Wolf creek, Coke Co. A. fruticosa var. *Caroliniana* S. Watson. Syn.

Robinia Pseudacacia L. Rich soil, O. S. April–May.

R. hispida L. Lookout Mts., Chattanooga. Specimens flowering when only a span high are found near summit. (R. hisp. var. *nana* Elliott.) May.

Wistaria frutescens DC. Low bottoms and banks of Cumberland river, Cockrill's bend. April–May.

Tephrosia Virginiana Pers. Dry, siliceous soils; oak barrens; common. June.

T. spicata Torr.& Gray. In same localities with the former. June–July.

Astragalus Canadensis L. Cliffs along Cumberland and other streams. June–July.

A. caryocarpus Kerr. Cedar glades near station at Lavergne. April.

A. Plattensis Nutt. var. **Tennessiensis** Gray. Very common in the limestone region of Middle Tenn. April–May.

Stylosanthes elatior Swartz. Argillaceous and siliceous soils. Harpeth hills, East Tenn; frequent. August.

Desmodium acuminatum DC. Rich woodlands. July.

D. Canadense DC. Cumberland Mts.; apparently rare in Tennessee. Cowan. July.

D. canescens DC. Highlands of Middle Tenn. July–August.

D. ciliare DC. Barrens. July–August.

D. cuspidatum Hook. Siliceous formation. Ridgetop. July–August.

D. Dillenii Darl. Very common. July–August.

D. humifusum Beck. Greenbrier, Ridgetop, etc. Highlands. July–August.

D. lævigatum DC. With the former. Harpeth hills. July–August.

D. Marilandicum Boott. Cedar and oak barrens. August–September.

D. nudiflorum DC. Rich woods. July–August.

D. paniculatum DC. Perhaps the most common species in limestone and siliceous soils. August.

D. pauciflorum DC. Deep rich soils in shady grounds. Harpeth hills, etc. July–August.

D. rigidum DC. Cedar glades, Lavergne. August.

D. rotundifolium DC. Rocky woodlands, hills near Nashville. August.
Var. **glabratum** Gray. Paradise ridge. August.

D. sessilifolium Torr. & Gray. Along railroad, Mitchellville, Sumner Co., etc. August.

D. viridiflorum Beck. Siliceous formation. Kingston Spring, South Tunnel, etc. August.

Lespedeza repens Bart. Argillaceous soils.
Var. **procumbens** Michx. In rocky cedar glades ; appears to be a distinct variety. July.

L. violacea Pers. With the former. July–August.

Lespedeza reticulata Persh. Barrens of Middle Tenn., Harpeth hills. August.

L. reticulata, var. angustifolia Maxim. Cedar glades. July–August.

L. stuvei Nutt. Highlands of Middle Tenn. August–September.

L. capitata Michx. Oak barrens at Tullahoma. July–August.

L. hirta Ell. Siliceous formation. July–August.

L. striata Hook. Over the whole State. Strictly avoiding limestone soils. Occurs in the remotest parts of the country and has been collected by me thirty years ago in East Tenn. It virtually carpets the ground.

Vicia Americana Mühlb. Thickets, vicinity of Nashville, etc. May.

V. micrantha Nutt. Copses and woods. April–May.

V. Caroliniana Walt. Highlands of Middle Tenn., but rare. Common in East Tenn. May.

Rhynchosia tomentosa Hook. Dickson Co. Frequent in the gravelly hills of lower East Tenn. July.

Phaseolus pauciflorus Benth. Imperfect specimens from near lunatic asylum, Nashville.

Apios tuberosa Mœnch. Bottom lands in moist thickets. August–September.

Phaseolus diversifolius Persoon. Cumberland Mts. June–September.

Ph. helvolus L. Very common in the barrens. June–September.

Ph. perennis Walt. Brownsville, West Tenn. August.

Clitoria Mariana L.   Margin of thickets and in barrens. June–July.

Centrosema Virginiana Benth.   Dry, open glades and barrens. June–September.

Amphicarpæa monoica Nutt.   Rich, damp woodlands.   August–September.

Galactia mollis Michx.   Common in all glades and barrens. May–September.

G. pilosa Elliott.   White bluff, Dickson Co.   Not rare.   July.

Baptisia australis R. Br.   Cedar glades at Lavergne.   June–July.

B. alba R. Br.   Cameron hill, Chattanooga; Mts. of East Tenn.; Ducktown.   June–July.

B. tinctoria R. Br.   Mts. of East Tenn.   Lookout Mt.   July.

Thermopsis fraxinifolia M. A. Curtis.   Region of copper mines, East Tenn.   May–June.

*Th. Caroliniana** M. A. Curtis.   Mountains along Ocoe river, East Tenn., and also frequent on summit of Harpeth ridge and Paradise ridge.   May.

Cladrastis tinctoria Raf.   Hills south of Nashville.   Trees attaining three feet diameter, but never otherwise than hollow. May–June.

Cercis Canadensis L.   Rich hillsides.   March–April.

Cassia Marylandica L.   Bottom lands and fence-rows.   July–August.

C. Tora L.   River bottoms, low marshy grounds.   July.

C. Chamæcrista C.   Sunny exposures in sandy soil.   July–August.

---

* *Thermopsis Caroliniana* M. Curtis, is a striking instance of distribution of species in widely distant colonies.   I first noticed the plant in the central range of the Big Frog Mts., near the copper mines at Ducktown, East Tenn., in 1860, in soils derived from quartzites and black roofing slates, in numbers, but not generally disseminated over the region.   Again I came upon it, in 1870, on the top of the ridges which encircle Nashville. These ridges are capped with siliceous cherts, which rest on the black Devonian shale.

C. nictitans L.   With the former.   July–August.

Gymnocladus Canadensis Lam.   Hill-sides in rich soil, vicinity of Nashville; also in Upper East Tenn.   April.

Gleditschia triacanthos L.   Over the State, especially in the glades.   June.

G. monosperma Walt.   In the cypress swamps and along streams in West Tenn.   July.

Schrankia angustata Torr. & Gray.   Open, dry glades and hillsides, East and Middle Tenn.   June–August.

Sch. uninata Willd.   Near Brownsville, West Tenn.   August.

Desmanthus brachylobus Benth.   Very frequent in the cedar glades.   July.

ROSACEÆ.

Prunus Americana L.   Shrub or tree, common in rich and poor lands.   March.

P. Chickasaw Michx.   Forming small thickets of circular outline, oldest in the center, young ones on the outskirts; also in fence-corners, etc.   March.

P. Chickasaw–Americana.   Popularly known as "Wild Goose Plum."   Frequent in Middle Tenn.

P. spinosa L.   Apparently introduced and cultivated on Hillsborough pike, twelve miles west of Nashville.   April.

P. Virginiana L.   Mountains of East Tenn., especially on the higher ones.   May.

P. serotina Ehrh.   In rich uplands over the State.   June.

P Pennsylvanica L.   Mountainous districts and high ridges of East Tenn.   On Clingman-Dom, 5000', occurs a variety with narrow lanceolate leaves.   May–June.

P. Persica Benth. & Hooker.   The peach is often found wild in woods and hedges.   March–April.

Neillia opulifolia L.   Rocky river banks, mouth of Mill creek, near Nashville.   May.

Spiræa Aruncus L.   Rich, moist woodlands here and there.   June.

S. tomentosa L.   In swampy regions of the highlands (Fountainhead, Sumner Co.).   July.

Gillenia trifoliata Moench. Rich woods, especially in the eastern part of the State. July.

G. stipulacea Nutt. Rich woods, hills vicinity of Nashville. June–July.

Agrimonia Eupatoria L. Common. July–September.

A. parviflora Ait. Prevalent in Middle Tenn., and very abundant. July–August.

Geum album Smel. Borders of woods over the State. May–July–August.

G. vernum Torr. & Gray. Common. March–April.

G. geniculatum Michx. High Mts. of East Tenn. Roane Mt. Prof. Chickering. August.

G. radiatum Michx. Roane Mt. Prof. Chickering.

Waldsteinia fragarioides Tratt. Mountains of East Tenn., Ocœ valley. June–July.

Potentilla Canadensis L. Dry, barren fields. May–July.

P. supina L. Near Hickman and Johnsonville, West Tenn. Moist ground. July–August.

P. tridentata Ait. Summit of Little Frog Mt., East Tenn.

Fragaria Virginiana Duchesne. Borders of woods, etc. March–April.
var. **Illinœnsis** Gray. In the cedar glades of Middle Tenn. April–May.

F. vesca L. Open woodlands and rocky places.

F. Indica L. Adv. Old graveyard, Nashville. May–June.

Rubus odoratus L. Higher Mts. of East Tenn. and the adjoining gorges; along Ocoe river, above Parksville; Big Frog Mt. June.

R. Canadensis L. Over the State, but more frequently in the Cumberland and Alleghany Mts. May.

R. occidentalis L. Hillsides and woodlands. July.

R. strigosus Michx. Woods and copses. More common around Nashville than the former. June.

R. villosus Ait. Fence-rows and waste grounds. June.
var. **frondosus** Gray and
var. **humifusus** Gray are frequently met with; also the

variety with white fruit occurs here and there.   (Cleveland, East Tenn.)

R. cuneifolius Pursh.   In sandy and siliceous soils.

R. trivialis Michx.   Sandy soils, common.   April–May.

R. hispidus L.   Wet and swampy localities; highlands and Mts. June.

Rosa Carolina L.   Low grounds and river swamps; Nashville, etc. June–July.

R. humilis Marsh.   Rocky river banks and cedar glades. May.

R. setigera Michx., var. *tomentosa* Torr. & Gray.   Abundant in the barrens of Middle Tenn. May.

R. rubiginosa L.   Roadsides and old fields. June.

R. bracteata Wendel.   Old homesteads and hedges; Nolinsville pike, six miles from Nashville. June–July.

R. pimpinellifolia L.   Introduced by early settlers; confined to old homesteads. July. (Col. Prosser's farm).

R. canina L. is credited to East Tenn., but has never been found by me.

Crataegus coccinea L.   From the summit of the Smoky Mts. to the lowlands of West Tenn. April.

C. cordata Ait.   Rocky glades around Nashville. May.

C. Crus Galli L.   Common. May.
  var. **ovalifolia** Lindl.   With the former. May.

C. flava Ait., var. **pubescens** Gray.   Copses. Middle Tenn. April.

C. Pyracantha Pers.   Low meadow near Hydt's ferry. May.

C. subvillosa Schrader.   Low tree or shrub, attaining a trunk of diameter of 18'.   Vicinity of Nashville, in rich soil. April.

C. tomentosa L.
  var. **pyrifolia** Gray.   Thickets around Nashville.
  var. **punctata** Jacq.   With the former. April–May.

Pyrus Americana DC.   Summit of Smoky Mts. Clingman-Dom, 6000'. July.

P. sambucifolia Chevn. & Schlecht.   Cultivated in a garden in Winchester, Franklin Co.   Not in the mountains!

P. angustifolia Ait.   Middle and East Tenn. Bon Air, Dickson Co. April.

P. arbutifolia L.   Mountain bogs, Cumberland and Alleghanies. May.

var. *melanocarpa* Elliot.   Laurel thickets, Sewanee. May.

P. Coronaria L.   Harpeth hills, near Judge John M. Lee's farm, at the foot of the hills.   March–April.

Amelanchier Canadensis Torr. & Gray.   From the high mountains to the valleys of Tennessee river.   April.

var. **Botryapium** Willd.   Swamps of West and Middle Tenn. April–May.

SAXIFRAGACEÆ.

Ribes Cynosbati L.   At the edge of a cedar glade near Forsterville.

R. rotundifolium Michx.   Summit of Roane and Smoky Mts. June.

Itea Virginica L.   Mts. of East Tenn. and cypress swamps of West Tenn.   July.

Hydrangea arborescens L.   Bluffs along Cumberland and other streams.   June–July.

var. *cordata* Torr. & Gray.   Banks of Ocoe river, above Parksville, East Tenn.   July.

H. radiata Walter.   Cataract at Tullahoma ; Cumberland Mts., above Sewanee ; summit of Lookout Mt.   June–July.

Decumaria barbara L.   Rocks along Ocoe river.   June.

Philadelphus hirsutus Nutt.   Cliffs on Cumberland river; ascent of ridge at Baker's station.   May.

Parnassia Caroliniana Michx.   Cumberland and Alleghany Mts. May–June.

P. asarifolia Vent.   Big Frog Mt., East Tenn.

Astilbe decandra Don.   Damp woods, especially in the mountains.   July.

Saxifraga Virginiensis Michx.   From the mountains to West Tenn.   March–April.

S. Careyana Gray.   Roane Mt.   Prof. Chickering, Wm. Canby.   July–August.

S. erosa Pursh.   Bed of Wolf creek and adjoining Bench Mt. ; Roane Mt.   Chickering.

S. leucanthemifolia.   Roane Mt.   Chickering.   Big Frog Mt.

Boykinia aconitifolia Nutt. Smoky Mts. July.

Heuchera Americana L. Vicinity of Nashville. May.

H. villosa Michx. Mts. of East Tenn.; Cumberland Mts.; Cliffs on Cumberland and highlands of Middle Tenn. Varies in size of flowers, outline and pubescence of leaves.

Mitella diphylla L. Shady glens in moist ground, valleys and Mts. of Cumberland and Alleghanies. June.

Chrysosplenium Americanum Schwein. Moist, shady ground along Ocoe river; near Mundi's bluff. April.

CRASSULACEÆ.

Sedum Nevii Gray. On argillaceous shists and siliceous conglomerate, Parksville, Polk Co. May.

S. pulchellum Michx. Covering large tracts of open, rocky glades in Middle Tenn. May.

S. Rhadiola DC. Roane Mt. Chickering.

S. telephioides Michx. Mts. of East Tenn. Chickering.

S. ternatum Michx. Shady, rocky places over the State. April–May.

Diamorpha pusilla Nutt. High rocky places in the vicinity of Sewanee and on Lookout Mt. May.

Penthorum sedoides L. Ditches and pools. Common. July– October.

HAMAMELIACEÆ.

Hamamelis Virginica L. Shrub, flowering often in midwinter. On the summit of Big Thunderhead I found it a low tree of twenty-five feet high, with a trunk of eighteen inches diameter (6000').

Liquidambar Styraciflua L. Low grounds. May.

HALORAGEÆ.

Myriophyllum verticillatum L. Tullahoma creek, Tullahoma.

Proserpinaca palustris L. Swampy lands, ditches. May.

P. pectinacea Lam. Swamps in the oak barrens, ditches along railroad embankments, etc. May.

MELASTOMACEÆ.

Rhexia Virginica L. Low wet lands. Common. July.
R. Mariana L. Low grounds in the oak barrens. June–July.

LYTHRACEÆ.*

Rotala ramosior Kœhne (Ammannia humilis Michx). Borders of ditches and swamps. July–September.

Ammannia coccinea Rottbœll. With the former. Has been distributed by me as "Ammannia latifolia."

Peplis diandra Nutt. (*Ammannia Nuttallii* Gray). Swamps along rivers, etc. July–September.

Lythrum alatum Pursh. Borders of springs and rivulets, near Cleveland, East Tenn. July.

Cuphæa petiolata Kœhne. (*Cuph. viscosissima* Jacq.) August–September.

Decodon verticillatus Ell. (*Nesæa verticillata* H. B. K.). Waters of Barren fork of Caney fork, near Nicholson springs. August.

ONAGRACEÆ.

Circæa Lutetiana L. Moist woodlands. May–June.

C. alpina L. Summit of Smoky Mts.; Roane Mt.

Gaura biennis L. Cumberland Mts. July–August.

G. filipes Spach. Chattanooga, Dr. Engelmann.

Jussiæa decurrens DC. Moist meadows, etc. July–August.

J. repens L. Swamps in Cockrill's bend, below Nashville. June.

Epilobium coloratum Muhlb. Woodlands, common. July–September.

Œnothera biennis L. Waste fields. Varies greatly.
var. **grandiflora** is common around Nashville. June–September.

Œ. fruticosa L. O. S. Baker's station, on the ridge. July.
var. *hirsuta* Nutt. In the poor, cherty soils of East Tenn.; Cleveland, Bradley Co. July.

---

*I adopt the revision of Lythraceæ by E. Kœhne, vide *Bot. Gaz.*, vol. x. no. 5.

Œ. glauca Michx.　Lookout Mt., Chattanooga.　June.

Œ. pumila L.　Mts. of East Tenn.; valley of Ococ river. June.　Also at Tullahoma.

Œ. sinuata L.　Near Hydt's ferry, Nashville.　June.

Œ. speciosa Nutt.　Introduced on vacant lots in Nashville. June.

Œ. triloba Nutt.　Frequent on pasture lands around Nashville.　June-July.

Ludwigia alternifolia L.　Swampy places.　June–July.

L. hirtella Raf.　Oak barrens at Tullahoma; Paradise ridge. July.

L. linearis Walt.　Bogs, in the oak barrens and on the highlands.　June–September.

L. palustris Ell.　Ponds and ditches.　July–August.

L. polycarpa Short & Peter.　With the former.　June–August.

CUCURBITACEÆ.

Echinocystis lobata.　Torr. & Gray.　Tullahoma.　July.

Sicyos angulatus L.　Hedges and thickets.　July.

Melothria pendula L　Cedar glades and hillsides.　July–September.

Trianosperma Boykinii Roem.　Found by Prof. Lester F. Ward, August, 1877, on the banks of Cumberland, above Nashville.　I have never succeeded in finding the plant.

PASSIFLORACEÆ.

Passiflora lutea L.　Thickets, common.　June–September.

P. incarnata L.　Dry soil, cultivated grounds.　A bad weed. May–September.

CACTACEÆ.

Opuntia Rafinesquii Engelm.　Sunny exposures in rocky glades.　May–June.

FICOIDEÆ.

Mollugo verticillata L.　A weed in fields and on roadsides. May–September.

UMBELLIFERÆ.

Hydrocotyle ranunculoides L. Ditches and swamps along Cumberland river. May.

H. umbellata L. Ditches near Hydt's ferry. May.

Sanicula Canadensis L. O. S. More prevalent in East Tennessee. June–August.

S. Marylandica L. Moist woodlands. June–August.

Eryngium yuccæfolium Michx. Dry barrens O. S., but nowhere numerous. July.

E. prostratum Nutt. Woods near Brownsville, West Tenn. August.

Daucus Carota L. Introduced and rapidly spreading. It grows more luxuriantly in East Tennessee than at home in Europe. June–July.

Polytænia Nuttallii DC. Paradise ridge. May.

Heracleum lanatum Michx. Big Frog Mt., Polk Co., East Tenn., 4000' alt. June.

Pastinaca sativa L. Apparently escaped from market gardens; not met with otherwise.

Archemora rigida DC. Rich woodlands. August.

Angelica Curtisii Buckl. Roane Mt. Chickering.

Archangelica hirsuta Torr. & Gray. Dry barrens; common. June–July.

Coriandrum sativum L. Introduced. Around dwellings and gardens, Nashville. June–July.

Ligusticum actæifolium Michx. Cumberland Mts., Lookout Mt., Roane Mt. June–July.

Thaspium barbinode Nutt. River banks and thickets. May–June.

Th. aureum Nutt. Woodlands. Common. June.

Th. trifoliatum Gray. With the former. June–July.

Zizia integerrima DC. Mountains of East Tenn., especially the higher ones. June–July.

Bupleurum rotundifolium L. Dry hills contiguous to the city of Nashville (Fort Naigele); also in the cedar glades on the Lebanon pike. May–June.

Discopleura capillacea Nutt. Some specimens on the grounds of the Motgomery-Bell Academy. Introduced.

D. Nuttalli DC. Damp woods vicinity of Cleveland, East Tenn. July.

Cicuta maculata L. Along streams. July.

Cryptotænia Canadensis DC. Common in low, damp woods. June–September.

Chaerophyllum Tainturieri Hooker. Very abundant in Middle Tenn. April–May.

Ch. procumbens L. Rich, moist woods, Nashville. April.

Osmorrhiza longistylis DC. Rich woodlands May–June.

Eulophus Americanus Nutt. Thickets along Charlotte pike; also on Murfreesborough pike. May–June.

Erigenia bulbosa Nutt. Rich woodlands. Common. March–April.

ARALIACEÆ.

Aralia spinosa L. Very copious in the glades of Middle Tenn. July.

A. racemosa L. Highlands of Middle Tenn.; hills along Charlotte pike, twelve miles north of Nashville. June.

A. hispida Michx. Mountains of East Tenn.; Ducktown. June.

A. nudicaulis L. Cumberland Mts., near Sewanee. June–July.

A. quinquefolia Decaisne & Planchon. Rich woods over the State, but everywhere rare. Harpeth hills, near Nashville. July.

CORNACEÆ.

Cornus florida L. In all localities. March–April.

C. alternifolia L. East Tenn.

C. stolonifera Michx. Banks of rivers and streams. May.

C. asperifolia Michx. Copses and dry ground. May–June.

C. sericea L. Wet places. Common. May.

Nyssa Caroliniana Poir. Along mountain streams, East Tenn.; Parksville; also Hollow Rock, West Tenn. May.

N. multiflora Wanger. O. S., especially in the barrens. April–May.

N. uniflora Walt. Swampy lands on Paradise ridge, West Tenn. April.

CARPRIFOLIACEÆ.

Sambucus Canadensis L. Waste places.

Viburnum Opulus L. Deserted homesteads. The cultivated variety only.

V. acerifolium L. Cumberland and Alleghany Mts. May–June.

V. lantanoides Michx. Mountains of East Tenn. July.

V. nudum L. Swampy lands in the barrens and in the mountains. May–June.

V. prunifolium L. Dry ground. Small tree. May.

Triosteum perfoliatum L. Foot of the mountains at Cowan. May–June.

Symphoricarpus vulgaris Michx. Rocky ground. A low bush. July.

Lonicera sempervirens L. Woods and barrens, climbing high. June–August.

L. Sullivantii Gray. Mountains of East Tenn. Vide Gray Flora, vol. 1, 2, p. 17.

Diervilla trifida Mænch. Cumberland and Alleghany Mts. June–August.

D. sessilifolia Benkley. Summit of Lookout Mt., along brooklets; Lula falls. June–July.

RUBIACEÆ.

Houstonia cœrulea L. Moist, open ground. April–August.

H. serpyllifolia Michx. Covering the ground in the high mountains. July.

H. patens Ell. Cedar glades, Lavergne. April.

H. purpurea L. Woodlands and copses. April–May.
  Var. **longifolia** Gray. Barrens
  Var. *tenuifolia* Gray. Mountains of East Tenn. July.

H. angustifolia Michx. Cedar glades; bluffs on Cumberland river. May–July.

Oldenlandia Boscii Chapm. Swamps near Tullahoma. July.

Cephalanthus occidentalis L. Wet places. Common. July–August.

Mitchella repens L. Dry woodlands. June–July.

Spermacoce glabra Michx. River banks. Very common. August.

Diodia Virginica L. Moist meadows; river banks; very frequent about Nashville. June–July.

D. teres Walt. Sterile grounds, old fields. Abounds around Nashville. July.

Galium Aparine L. Waste grounds. April–May.

G. virgatum Nutt. Dry, sterile places in the cedar glades, Lavergne. June.

G. pilosum Ait. Dry copses. Common. June–July.

G. trifidum L. Swampy meadow lands. July–October.

G. triflorum Michx. Dry woodlands; copiously in the cedar barrens. July.

G. circæzans Michx. Hills around Nashville to the mountains of East Tenn. June–July.

G. latifolium Michx. High mountains of East Tenn.; Big Frog Mt. July.

G. lanceolatum Torr. East Tenn.

G. Arkansanum Gray. Johnsonville, West Tenn.

VALERIANACEÆ.

Valerianella radiata Dufr. Glades and copses. April–May.

V. Woodsiana Walp.
  Var. **umbilicata** Gray. Glades and pastures. April–May.
  Var. **patellaria** Gray. With the former.

DIPSACEÆ.

Dipsacus sylvestris Mill. Waste grounds. Common. July.

COMPOSITÆ.

Elephantopus Carolinianus Willd. Woods. June–July.

Vernonia Novæboracensis Willd. Roadsides and open woodlands. August–September.

V. Baldwinii Torr. West Tenn. August.

V. altissima Nutt. River bottoms. August.

V. fasciculata Michx. With the former. Very common. August.

Eupatorium purpureum L. Low grounds. Common. August–September.

E. serotinum Michx. Waste places in country and towns. September–October.

E. album L. Paradise ridge, Robertson Co., East Tenn. July–August.

E. semiserratum DC. (E. parviflorum Elliott). Barrens at Tullahoma. July–August.

E. altissimum L. Dry copses. Copiously on Grany White pike, near Mrs. Cheatham's place. August–September.

E. rotundifolium L. Mitchellville, Tullahoma. July–August.

E. leucolepis Torr. & Gray. Barrens of Middle Tenn. August–September.

E. perfoliatum L. Low grounds. July–September.

E. incarnatum Walt. Glades and copses. August–September.

E. ageratoides L. Rich woodlands. Over the State. August–September.

E. aromaticum L. Barrens at Tullahoma. July.

E. cœlestinum L. Moist grounds. Common. (Conoclinum cœlestinum DC.) September.

Mikania scandens Willd. Johnsonville. West Tenn. September.

Kuhnia eupatorioides L. Common. September.
Var. **corymbulosa** Torr. & Gray. Vicinity of Nashville. September.

Liatris squarrosa Willd. Highlands. Greenbrier, Robertson Co. June–July.

L. scariosa Willd. Mountains of East Tenn. Lookout Mt. July.

4

L. spicata Willd. Oak barrens of Middle Tenn. to mountains of East Tenn. July.

L. graminifolia Pursh. With the former. July.

Grindelia lanceolata Nutt. Cedar glades at Lavergne, near the station. July.

Chrysopsis graminifolia Nutt. Mountain regions of East Tenn. July.

Ch. Mariana Nutt. In sandy soil. Common. June–July.

Solidago cæsia L. Harpeth hills, near Nashville. September.

Var. **paniculata** Gray. With the former. September.

S. latifolia L. Rocky banks of Richland creek, near Nashville. August–September.

S. Curtissii, var. **pubens**. Roane Mt. J. W. Chickering. August.

S. monticola Torr. & Gray. Clingman-Dom of the Smoky Mts. Roane Mt. J. W. Chickering.

S. bicolor L. Kingston Springs, Dixon Co., East Tenn. August–September.

S. glomerata Michx. Clingman-Dom, 6000'; Roane Mt. J. W. Chickering.

S. spithamea, M. A. Curtis; Roane Mt., J. W. Chickering. August.

S. odora Ait. Barrens and highlands. July.

S. speciosa Nutt. Highlands, Greenbrier; glades at Lavergne. August–September.

Var. **angustata** Torr. & Gray. Common in siliceous soils. August–September.

S. rugosa Mill. Highlands and barrens. August–September.

S. ulmifolia Mühlb. With the former. August–September.

S. arguta Ait. Lookout Mt.; Ocoe region. August–September.

S. juncea Ait. Oak barrens and woodlands. August–September.

*8. Gattiugeri Chapm. no. sp. In the cedar glades, near railroad station at Lavergne, Rutherford Co. August–September.

S. serotina Ait. Fields and fence-rows. September. Var. gigantea Gray. With the former.

S. Canadensis L. Fields and river banks. Very abundant. September. Var. procera Torr. & Gray. Low, moist meadow lands. September–October.

S. patula Mühlb. Near Cranberry Iron Works, East Tenn. September. Mrs. E. I. Britton.

S. Shortii Torr & Gray. Barrens at Tullahoma. Abundant, with S. juncea. August. Also Bon Air, Dickson Co.

S. nemoralis Ait. Abundant in all glades and barrens. August–September.

S. corymbosa Ell. Bon Aqua station, Dickson Co., Tullahoma. August–September.

S. lanceolata L. Highlands. Fountainhead, Robertson Co. August.

S. rupestris Raf. Cliffs facing Cumberland river. August.

Brachychæta cordata Torr. & Gray. O. S. September.

Bellis integrifolia Michx. Copses and barrens. July.

Boltonia asteroides L'Her. In a low meadow near Hydt's Ferry, Nashville. September.

B. diffusa Elliott. In wet, sandy soil. Paradise ridge, Tullahoma. July.

---

* *Solidago Gattingeri* Chapm. ined. Slender, upright, 2–4 feet high : branches and inflorescence perfectly smooth and glabrous: leaves ciliolate; lower cauline and radical lanceolate-spatulate, appressed serrulate, obviously tripli-nerved; upper cauline mainly entire and without lateral ribs, oblong lanceolate and an inch or so long, and the upper reduced to half or quarter inch, but near the inflorescence very small and bract-like; racemiform clusters of small heads open and spreading, not recurving, disposed to forming a corymbiform very naked panicle : involucral bracts oblong, very obtuse, yellowish in the dried plant: flowers 15–20 in the head. rays 4–6 : akenes appressed-puberulent or the lower part glabrous. *S. Missouriensis*, var. *pumila* Chapm. Fl. Suppl. 627. Between *Missouriensis* and *Shortii* Gray. It occurs in numerous individuals over a couple of acres and is not likely to be a hybrid. The associated species are S. nemorosa (very abundant), S. speciosa, var. angustata, S. speciosa, S. Canadensis. First collected September, 1869.

Sericocarpus conyzoides Nees.　Dry, argillaceous soils.　July.

S. solidagineus Nees.　With the former.　July.

Aster paludosus Ait.　Moist ground in the barrens.　August–September.

A. corymbosus Ait.　Mountain districts.　Sewanee.　July.

A. Curtisii Torr. & Gray.　Roane Mt.　J. W. Chickering.　September.

A. oblongifolius Nutt.　Banks of Cumberland river.　Bluffs along Mill creek, near Nashville.　August.

A. concolor L.　Barrens at Tullahoma.　September.

A. patens Ait.　Highlands.　September.
Var. **gracilis** Hooker.　With the former.　September.

A. Shortii Hooker.　Rich woodlands in the hills around Nashville.　September.

A. undulatus L.　With the former.　August–September.　Lookout Mt.

A. cordifolius L.　Rocky banks on Cumberland river and its tributaries.　September.

A. Drummondii Sindl.　Vicinity of Nashville.　September.

A. sagittifolius Willd.　Wild goose pond, near Mitchellville.　September.

A. macrophyllus L.　Cumberland Mts., near Tracy City.　July.

A. lævis L.　Rocky river banks.　Nashville.　September.

A. ericoides L.　Covering old fields and waste places.　September–October.
Var. **Rivesii** Gray.　A handsome variety, with larger purple flowers.　Rocky river banks and glades, near Nashville.　September.
Var. **villosus** Torr. & Gray.　Occurs with the former on dry, rocky ground.

A. multiflorus Ait.　Argillaceous and silaceous soils.　Highlands.　September–October.

A. dumosus L.　With the preceding.　September.

A. diffusus Ait.　A very variable species, growing in great abundance on muddy river banks and bottoms.　August–September.
Var. **horizontalis** Gray.　Thickets along Cumberland.　September.

Var. thyrsoideus Gray. Pastures; open grounds. Hydt's Ferry, etc.

Var. bifrons Gray. With much larger heads, and long lanceolate acuminate foliage. Shady banks of Cumberland river. September–October.

A. Tradescanti L. Replaces in the siliceous soils aster ericoides in equal abundance. September–October.

A. umbellatus Mill. Highlands, Middle Tenn. September.

A. infirmus Michx. (*Diplopappus cornifol.* Darl.) Chilhowee Mts., East Tenn. September. A. H. Curtiss.

A. linariifolius L. Mountains of East Tenn. Also highlands of Middle Tenn. Harpeth Hills, near Nashville. September.

Erigeron Canadensis L. Common weed. August–September.

E. divaricatus Michx. Sandy barrens, Cedar Hill, Robertson Co. August.

E. bellidifolius Mühlb. Highlands, and East Tenn. April–May.

E. Philadelphicus L. In meadows. May.

E. annuus Pers. Dry glades. May–June.

E. strigosus Mühlb. Pastures and waste ground. September–October.

Plinthea camphorata DC. Low, damp grounds. O. S. July–September.

Var. *Beyrishii* Torr. & Gray. West Tenn.

Antennaria plantaginifolia Hawk. Dry copses and hillsides. Common. June.

Gnaphalium polycephalum Michx. Open woodlands, etc. July–September.

G. decurrens Ives. Highlands. June–August.

G. uliginosum L. Swamps along Cumberland Mts., etc.

G. purpureum L. Common. September.

Polymnia Canadensis L. Deep ravines in the Cumberland Mts.; Sewanee. Abundant. July.

Var. radiata Gray looks very different and should perhaps be counted as a species. Forsterville, at the edge of a cedar barren; Cockrill's bend, in a dry barren; near Nashville. July.

P. Uvedalia L. Rich woodlands. Abundant about Nashville. July–August.

Silphium perfoliatum L. On Brown's creek, three miles from Nashville. August.

*S. brachiatum Gattinger. First collected July, 1867, on the foot of Cumberland Mts., near Cowan, Tenn.

S. integrifolium Michx. Highlands: Charlotte pike; Davidson's place, near Nashville. July.

S. scaberrimum Elliott. Barrens at Tullahoma. July.

S. Asteriscus L. Craggie Hope, Cheatham Co., near railroad. July.

Var. *lævicaule* DC. Barrens at Tullahoma. August.

S. trifoliatum L. Copses around Nashville. Frequent. July.

S. compositum Michx. Western declivities of Chilhowee Mts. and in the hills on Chestua, East Tenn. July.

S. terebinthinaceum Jacq., var. *pinnatifidum* Gray. Barrens at Lavergne, Tenn. June–July.

S. laciniatum L. Apparently rare in this State. East Tenn.

Chrysogonum Virginianum L. Thickets in the mountains at Ducktown. April.

Parthenium integrifolium L. Dry ground. Common. July.

Inula Helenium L. Sweetwater East Tenn. July.

Ambrosia bidentata Michx. Fields and pastures. Brownsville. West Tenn. August.

A. trifida L. Banks of streams and bottom lands. August.

Var. **integrifolia** Torr. & Gray. Is only a depauperate form, always in very poor soil.

---

* *Silphium brachiatum* Gattinger, n. sp. Stem 3–5 feet high, square or subangular, with the brachiated, thin, roundish and nearly leafless flowering branches smooth and glaucous. Leaves opposite, roughened on the upper side, smooth on the lower, except the principal veins, which are slightly hirsute, lower short petioled, deltoid or hastate-lanceolate, irregularly and upwardly dentate, 6–10 inches long, green and glaucous, those on the branches distant, small, sessile, entire; heads long peduncled. small, bracts of the involucre ovate; achenia obovate-orbiculate. narrow winged, slightly notched at the apex. Very distinct. The 1–3 flowered peduncles 3–4 inches long and almost filiform. Involucre little over one-half inch high. Rays rather few, one-half inch long; akenes four lines long.
Collected July 14th, 1867, on the western slope of the Cumberland mountains, a short distance south of the tunnel at Cowan, Tennessee. It is quite numerous on the limestone base of the mountain, and probably extends all along the Chattanooga road towards the Tennessee river. It was, however, not since seen in any other part of the State.

A. artemisiæfolia L. Fields and cultivated grounds. The most abundant weed in this region. July.

Xanthium Canadense Mill. In rich bottoms it attains sometimes six feet in height. July.

X. spinosum L. Towns and railroad depots. Nashville, Knoxville. July.

Zinnia pauciflora L. Escaped from gardens. Near Charlotte pike. July.

Heliopsis lævis Pers. Highlands and mountains of East Tenn. July:

Ecclipta alba Hasskarl. River banks and miry places. Common. July–September.

Echinacea purpurea Moench. Rich woodlands. More frequent in East Tenn. Whiteside. June–July.

E. angustifolia DC. Dry copses, vicinity of Nashville; Cedar glade near Lavergne. July.

Rudbeckia triloba L. Thickets. July–August.
  Var. *rupestris* Chickering. Roane Mt. Chickering.

R. hirta L. Dry copses and barrens. August–September.

R. fulgida Ait. Dry woodlands. September.

R. speciosa Wendeworth. In the glades. Lavergne. September.

R. spathulata Michx. Chilhowee Mt., East Tenn. A. H. Curtiss. September.

R. bicolor Nutt. Open, dry barrens, Nashville. Bon Air. September.

R. laciniata L. Paradise ridge. Mountains of East Tenn. (Wolf creek.) July.

Lepachys pinnata Torr. & Gray. Rocky banks and rich pasture lands. July.

L. columnaris Torr. & Gray, var. *pulcherrima* Torr. & Gray. Vicinity of a cotton compress, Nashville. Introduced. July–August.

Helianthus annus L. Escaped from gardens. Not indigenous in Tenn. July–August.

H. strumosus L. . Cumberland Mt. at Cowan. July.

H. atrorubens L.   Dry woodlands in siliceous soil.   East and Middle Tenn.   July.

H. Schweinitzii Torr. & Gray.   Cowan.   August.

H. mollis Lam.   Barrens of Middle Tenn.   Frequent.   July–August.

H. parviflorus Bernh.   Highlands.   July–August.

H. divaricatus L.   Highlands and argillaceous soils generally. July–August.

H. hirsutus Raf.   Barrens and woodlands.   Common.   July–August.

H. decapetalus L.   Mountains of East Tenn.; Paradise ridge.   July.

H. tuberosus L.   Rich bottom lands.   Very frequent around Nashville.   July–August.

H. lætiflorus Pers.   Brownsville, West Tenn.   August–September.

H. tracheliifolius Willd.   Mountains of East Tenn.; Big Frog Mt.   July.

Helianthella tenuifolia Torr. & Gray.   Barrens, two miles east of Tullahoma.   July–August.

Verbesina occidentalis Walter.   Western declivities of Smoky Mts., covering large tracts.   July–August.

V. Virginica L.   Limestone regions of Middle Tenn.   August–September.

V. helianthoides Michx.   Rich woodlands.   August–September.

Actinomeris squarrosa Nutt.   Moist woodlands.   August–September.

Coreopsis rosea Nutt.   In a little swamp on Potato creek, Copper hill; Ducktown.   July.

C. auriculata L.   Paradise ridge; Greenbrier, Robertson Co. June.

C. delphinifolia Lam.   East Tenn.

C. senifolia Michx.   In siliceous soil (gravelly ridges), over the State.   July–August.

    Var. **stellata** Torr & Gray.   With the former.

C. tripteris L.   Very abundant.   August–September.

C. verticillata L. Upper East Tenn. August–September.

C. trichosperma Michx. River swamps, Nashville. August–September.

C. discoidea Torr. & Gray. River swamps, with the former. August–September.

Bidens frondosa L. Moist, low grounds. July–October.

B. connata Mühlb. July–October.

B. cernua L. Abundant in wet grounds. July–September.

B. bipinnata L. Cultivated grounds and barrens. July–August.

Marshallia lanceolata Pursh. Vicinity of Memphis. Dr. G. Egeling.

Galinsoga parviflora Cav. Cultivated grounds, Mrs. Cheatham's place, Nashville. September–October.

Helenium nudiflorum Nutt. Wet, sandy soil; Paradise ridge, Tullahoma. July–August.

H. tenuifolium Nutt. Argillaceous, sandy soil, Brownsville, etc. July–August.

H. parviflorum Nutt. Cedar glades, Lavergne, etc. July.

H. autumnale L. Low, moist places. September.

Dysodia chrysanthemoides Lag. Along railroad embankments near Mitchellville. Abundant across the State line in Kentucky. August.

Anthemis Cotula L. Waste places. May.

Achillea Millefolium L. Roadsides and meadows. August.

Chrysanthemum Leucanthemum L. Roadsides and dry pastures. Frequent about Nashville.

Tanacetum vulgare L. Escaped from gardens. July–August.

Artemisia Ludoviciana Nutt. Roadsides. Infrequent. June–July.

A. annua L. Introduced. Now spreading over the State. August–September.

A. Absinthium L. Old homesteads and thence escaped into fence-rows. West Tenn. Dr. Chapmann.

A. biennis Willd. Very common on the banks of the Ohio at Louisville. West Tenn.

Senecio Rugelia Gray. Smoky Mts. Gray Synops., vol. i. 2 p. 383.

S. aureus L., var. **obovatus** Torr & Gray. Moist ground in limestone regions. May.

Var. **Balsamitæ** Torr & Gray. Mountains of East Tenn. Ducktown. May.

S. lobatus Pers. West Tenn., low grounds. June.

Cacalia suaveolens L. Banks of Turnbull creek, near Kingston Springs. July–August.

C. reniformis Mühlb. Cumberland and Alleghany Mts. July–August.

C. atriplicifolia L. Rich woodlands. July.

C. tuberosa Nutt. Wet places in the cedar glades. July–August.

Erechtites hieracifolia Raf. Clearings in forests. Common. July.

Arctium Lappa L. In towns and on highways. June–July.

Cnicus lanceolatus Hoffm. Waste places. July.

C. altissimus Willd., var. **filipendulus** Gray. Fence-rows near Lavergne. June–July.

Var. **discolor** Gray. Banks of Cumberland on Rising Sun Bluff, twelve miles below Nashville. September.

C. muticus Pursh. Lookout Mt., Chattanooga; also Roane Mt. Chickering. July.

Onopordon Acanthium L. Dry pastures around Nashville. July–August.

Silybum Marianum Gaert. One specimen found along N. W. Railroad, near Nashville, and foliage brought to me by Mrs. Turner, of Nashville, found near Tracy City.

Centaurea Cyanus L. Escapes sometimes from gardens.

Krigia Virginica Willd. Sandy soil. Sewanee, Lookout Mt and near Prospect Station, Giles Co. June.

K. Dandelion Nutt. Moist woodlands. April–May.

K. amplexicaulis Nutt. Rich woodlands. May.

K. montana Nutt. High mountains of East Tenn. Roane Mt. Chickering.

Cichorium Intybus L. Roadsides. Edgefield. Rare. July

Hieracium paniculatum L.   Dry, open woodlands.   July.

H. venosum L.   With the former.

H. scabrum Michx.   Rocky places.   July–August.

H. Gronovii L.   Frequent in the hills around Nashville.   July–August.

H. longipilum Torr.   Johnsonville.   July.

Tragopogon pratensis L.   Mrs. Cheatham's grounds, and vicinity.   Nashville.   May.

Preuanthes crepidinea Michx.   Rich soil.   Hills near Nashville.   August.

P. aspera Michx.   Barrens.   July–August.

P. serpentaria Pursh.   Barrens of Dixon Co.   September.
Var. *barbata* Gray.   Roane Mt.   Chickering.   Smoky Mts., Big Thunderhead.   July–August.

P. altissima L.   Rich woods; highlands.   July–September.

Taraxacum officinale Webr.   Cultivated lands, everywhere. April–September.

Pyrrhopappus Carolinianus DC.   Cleveland, East Tenn. West Tenn.   July.

Lactuca Canadensis L.   Clearings and woodlands.   June.

L. acuminata Gray.   Bluffs on Cumberland river.   September.

L. Floridana Gaert.   Rich woodlands.   August–September.

L. leucophæa Gray.   Thickets along Cumberland river, near Nashville, etc.   July.

Sonchus oleraceus L., and

S. asper L.   Both in cultivated ground.   Common.   June–July.

LOBELIACEÆ.

Lobelia cardinalis L.   Wet, miry ground.   July–September.

L. syphilitica L.   Swamps and ditches.   July–September.

L. puberula Michx.   Highlands.   August–September.

L. leptostachys A. DC.   Barrens in moist ground.   July–August.

L. spicata Lam.   Cedar glades.   With the former.   July.

*L. Gattingeri Gray. Cedar glades of Middle Tenn. Rocky banks of Cumberland river. Frequent May–June.

L. Nuttallii Roem. & Schult. Mountain swamps. Sewanee. July.

L. inflata L. Dry, argillaceous and siliceous soils. August–September.

CAMPANULACEÆ.

Specularia perfoliata A. DC. Waste ground and roadsides. April–May.

Campanula aparinoides Pursh. Mountain swamps and laurel thickets. July–August.

C. divaracata Michx. Cumberland and Alleghany Mts. July–August.

C. Americana L. Thickets and dry woodlands. August–September.

ERICACEÆ.

Gaylussacia brachycera Gray. Mountains on Ocoe river. June–July.

G. frondosa Torr. & Gray. Mountains of East Tenn. April–May.

G. resinosa Torr. & Gray. Highlands. Valley of East Tenn., etc. April–May.

Vaccinium arboreum Marsh. Highlands and siliceous soil generally. May.

V. staminenm L. Oak barrens and highlands. May–June.

V. corymbosum L. Mountains and ridges of East Tenn. Highlands. April–May.
Var. *pallidum* Gray. (*V. Constablei* Gray.) Mountains of East Tenn. June.

**Lobelia Gattingeri* Gray. Flowers 4–5 lines long, deep blue; stem smooth, weak and branching; leaves thin, sessile, oblong-ovate, obtuse, serrate, the lowest obovate; racemes peduncled, very slender, many-flowered; calyx-tube ovoid, longer than its pedicel, shorter than the linear-subulate entire lobes, the sinuses not appendaged, but slightly callous. The pedicels are sometimes provided with small bracteols. Plant 6–20 inches high. Regular flowering time first week in May, but some plants are found flowering as late as August. Its home is the limestone basin of Middle Tenn., and prominently moist places in the cedar glades First collected about 1869, at Lavergne, seventeen miles south of Nashville.

V. hirsutum Binkley.   High mountains of East Tenn.   June–July.

V. erythrocarpon Michx.   Higher mountains of East Tenn. June–July.

Epigœa repens L.   Cumberland and Alleghany Mts.   March–April.

Gaultheria procumbens L.   Alleghany Mts., throughout.   June.

Andromeda Mariana L.   Cumberland and Alleghany Mts. April–May.

A. ligustrina Mühlb.   Smoky Mts.   Big Frog Mts.   July.   ✗

Oxydendron arboreum DC.   Highlands and siliceous soils. April–May.

Leucothœ Catesbayi Gray.   Along streams throughout Cumberland and Alleghanies.   Common.   April–May.

L. recurva Gray.   With the former.   May.

Kalmia latifolia L.   Mountains of East Tenn., and in Middle Tenn., in siliceous soils.   May–June.

Menziesia globularis Salisb.   High mountains of East Tenn. Roane Mt.   Clingman Dome.   July.

Rhododendron arborescens Torr.   Big Frog Mts.   June.

R. viscosum Torr.   Along mountain streams.   Ocoe Valley, etc.   June.

R. nudiflorum Torr.   Hills of East and Middle Tenn., in siliceous soils.   April.

R. calendulaceum Torr.   Alleghany and Cumberland Mts.; also here and there in shaded ravines in the valleys.   Cave Spring, Roane Co.   June.

R. maximum L.   Common throughout Cumberland and Alleghany Mts.   June–July.

P. Catawbiense Michx.   Only on the highest summits of Smoky Mts. (Clingman Dome) and Roane Mt.   June–July.   ↖

Leiophyllum buxifolium Ell., var. *prostratum* Gray.   Summit of Roane Mt..   Chickering.   July–August.

Clethra acuminata Michx.   Throughout Alleghany Mts.   July.

Chimaphila maculata Pursh.   Over the State.   June–July.

Ch. umbellata Nutt.   Sewanee.   Tracy City.   July.

Pyrola elliptica Nutt.   Wolf Creek (Bench Mt.).   July.

Monotropa uniflora L. Hills near Nashville. August.

M. Hypopitys L. Oakdale Station, C. South. Railroad, near Wartburg. Cumberland Mts. July.

Galax aphylla L. Very frequent in the Alleghany Mts. June–July.

PRIMULACEÆ.

Dodecatheon Meadea L. The purple flowering variety in East Tenn. All specimens found in the vicinity of Nashville had white flowers. May–June.

Steironema ciliatum Raf. Common. June–August.

S. lanceolatum Gray. Over the State. June–July.
Var. *angustifolium* Gray. East Tenn., Bradley Co. July.

Lysimachia Fraseri Duby. Lookout Mt. Rocks along Ocoe river, Polk Co. June–July.

L. quadrifolia L. Woodlands. July.

L. Nunularia L. Escaped from gardens. July.

Anagallis arvensis L. Railroad embankments. Oakland Station, Robertson Co.

A. cœrulea L. Grass plots. Nashville. Introduced. June.

Samolus Valerandi L., var. *Americanus* Gray. Muddy river banks and wet places. May–July.

SAPOTACEÆ.

Bumelia lycioides Gaert. Shrub or small tree in rich soil. Middle Tenn. and lower part of East Tenn. June–July.

EBENACEÆ.

Diospyros Virginiana L. Common. Medium size tree. June.

STYRACEÆ.

Halesia tetraptera L. Ocoe district, East Tenn. March–April.

OLEACÆ.

Fraxinus Americana L. Over the State. April.
Var. **microcarpa** Gray. Harpeth hills, near Nashville. Supposed to be a hybrid between F. Americana and viridis.

F. viridis Mich. Low, moist ground. April.

F. pubescens Lam. Edge of river swamps. April.

F. quadrangulata Michx. Hillsides, in rich ground. April.

Forestiera acuminata Poir. Close at the water's edge. Cumberland river and tributaries. March.

F. ligustrina Poir. Characteristic shrub for the cedar glades of Middle Tenn. July.

Chionanthus Virginica L. Along streams, especially in the Alleghany Mts. April.

Ligustrum vulgare L. The common Privet, planted in hedges. Seldom perfects its fruits in Middle Tenn. May.

APOCYNACEÆ.

Amsonia Tabernæmontana Walt. (*A. latifolia* Michx., East. Tenn., and *Ams. salicifolia* Pursh., Nashville.) Over the State. May–June.

Vinca minor L. Introduced. Old graveyards, where it grows luxuriantly, perfecting seeds. Not spreading, however, into neighboring woodlands.

Apocynum androsæmifolium L. O. S. June–July.

A. cannabinum L. O. S. June–July.

ASCLEPIADACEÆ.

Asclepiodora viridis Gray. Cedar glades, in dry, sunny places. July–August.

Asclepias tuberosa L. Fields and pastures. June–July.

A. purpurascens L. Edges of woodlands and copses. June–July.

A. incarnata L. River swamps. June–July.

A. Cornuti Decaisne. Open grounds. July.

A. obtusifolia Michx. Ocoe district, East Tenn. June–July.

A. phytolaccoides Pursh. Mountains of East Tenn. July.

A. variegata L. Dry uplands. June–July. East Tenn., and vicinity of Nashville.

A. quadrifolia L. Oak barrens of Middle Tenn. June–July.

A. verticilliata L. Frequent in the cedar glades.

Acerates viridiflora Ell. Cedar glades, Lavergne. June–July.

A. longifolia Ell. Barrens, Tullahoma. July–August.

Enslenia albida Nutt. Thickets. Common. July.

Gonolobus lævis Michx. Copses vicinity of Nashville. East Tenn.

Var. **macrophyllus** Gray. With the former. July.

G. obliquus R. Br. Thickets. Copses near Nashville. June–July.

G. hirsutus Michx. Thickets along Cumberland river and Stoner's creek. June–July.

LOGANIACEÆ.

Gelseminum sempervirens Ait. Lookout Mt., East Tenn. Abundant in West Tenn. March–April.

Spigelia Marilandica L. Over the State. May–June.

Polypremum procumbens L. West Tenn. Brownsville. Dry pastures. April–September.

GENTIANEÆ.

Sabbathia brachyata Elliott. Barrens. June–July.

S. angularis Pursh. Pastures, in rich soil. July.

S. gracilis Pursh. Barrens at Tullahoma. July.

Gentiana quinqueflora Lam. Mountains around Ducktown. August–September.

G. Saponaria L. Moist thickets, highlands. September–October.

G. Andrewsii Grieseb. Highlands, South Tunnell. August–September.

G. ochroleuca Froebl. Barrens and highlands. September.

Frasera Carolineusis Walt. East Tenn. Corkills Bend, near Nashville. June.

Obolaria Virginica L. Moist woods north of South Tunnell, Sumner Co. April.

Bartonia tenella Mühlb. Mountain bogs. Sewanee. July.

POLEMONIACEÆ.

Phlox paniculata L. Over the State, in rich, moist woodlands. June–July.

P. maculata L. Similar localitites. July–September.

P. glaberrima L. Mountains of East Tenn. to Middle Tenn.
Var. **suffruticosa** Gray. Banks of Cumberland river at Nashville. June–July.

P. amœna Sims. Mountains of East Tenn. and highlands of Middle Tenn.; Mitchellville, etc. May–June.

P. divaricata L. Over the State. April–May.

P. Stellaria Gray. Cedar glades, Lavergne. May.

Gilia coronopifolia Pus. Knobs, east of Athens, East Tenn. Perhaps only escaped from cultivation. July.

Polemonium reptans L. Moist woodlands from the mountains to the Mississippi. April–May.

HYDROPHYLLACEÆ.

Nemophila microcalyx Fish. & Mayer. Thickets and ravines. Nashville. April.

Phacelia bipinnatifida Michx. Over the State. Shady localities. April–May.

P. Purshii Buckley. Very copiously distributed in Middle Tenn. April.

P. parviflora Pursh. Rocky glades, vicinity of Nashville. April.
Var. **hirsuta** Gray. With the former.

Hydrophyllum macrophyllum Nutt. Wet woodlands. South Tunnel, Sumner Co. June.

H. Canadense L. Mountains of East Tenn. July.

H. appendiculatum Michx. Frequent in vicinity of Nashville. Rich woodlands and river bluffs. June.

H. Virginicum L. Ducktown, East Tenn. June.

Hydrolea affinis Gray. Hollowrock. August.

BORRAGINACEÆ.

Heliotropium tenellum Torr. Cedar glades, Middle Tenn. Lavergne. July.

5

H. anchusæfolium Poir. Introduced. Grounds near Mrs. Cheatham's, Nashville. July.

H. Indicum L. Low, wet grounds. July–September.

Cynoglossum officinale L. Waste grounds. Over the State. May.

C. Virginicum L. Rich woodlands. Common. May.

Echinospermum Virginicum Lehm. Vile weed. Common. June–August.

E. Lappula Lehm. Johnsonville. July.

Mertensia Virginica DC. Rich, moist soil. Over the State. May.

Myosotis verna Nutt. Poor, rocky soil. Common. April.

Lithospermum arvense L. Waste places. April.

— **L. latifolium** Michx. Highlands; in moist woodlands. South Tenn., Sumner Co. June.

L. canescens Lehm. Glades of Middle Tenn. June–July.

L. angustifolium Michx. West Tenn., near Hickman. June.

Onosmodium Carolinianum DC., var. *molle* Gray. Vicinity of Nashville. April–May.

Symphytum officinale L. Introduced by early settlers in Bayer's settlement, Polk Co., East Tenn. July.

CONVOLVULACEÆ.

Ipomœa Ouamoclit L. Cultivated and frequently spontaneous. July–September.

I. coccinea L. Cultivated grounds; very common in cornfields. July–September.

I. hederacea Jacq. Fields and waste places. July–September.

I. purpurea Lam. Frequently cultivated and escaping. June–August.

I. pandurata Meyer. Copses and river banks. June–September.

I. lacunosa L. Fields and waste grounds. July–September.

Convolvulus spithamœus L. Dry, rocky woods frequent in the Alleghanies. May–September.

C. sepium L. Moist alluvial soil. Along Cumberland river. August–September.

C. arvensis L. Grass plots, Nashville and vicinity. June–July.

Evolvulus argenteus Pursh. Dry rocky places in the cedar glades at Lavergne. May–June.

Cuscuta chlorocarpa Engelm. Moist thickets on Mill creek, etc. August–September.

C. arvensis Beyrich. Open grounds in the glades on Ambrosia. June–July.

C. tenuiflora Engelm. Mountains of East Tenn.; Big Frog Mt., Polk Co. July.

C. Gronovii Willd. On shrubs and herbaceous plants, over the State. August–September.

C. rostrata Shuttleworth. Summit of Thunderhead, on Solidago glomerata. July.

C. compacta Juss. On Eupatorium and other herbaceous plants. Paradise ridge. August–September.

C. glomerata Choisy. On low bushes, highlands. July–September.

SOLANACEÆ.

Solanum nigrum L. Cultivated and uncultivated grounds. May–September.

S. Dulcamara L. Escaped from cultivation. Jellico, East Tenn., etc. June–July.

S. Carolinense L. Fields and gardens. May–July.

S. rostratum Dun. Introduced in Nashville and vicinity and spreading. July–August.

Physalis angulata L. Copses and pastures. July.

P. Philadelphica Lam. Fields and cultivated grounds. July.

P. Virginiana Mill. Cedar glades and woods. Middle Tenn. June–September.

P. pubescens L. Sandy banks of rivers. July–September.

Nicandra physaloides Gaertn. Fields and gardens. August–September.

Lycium vulgare Dun. Roadsides and waste places. July–September.

Datura Stramonium L. Waste ground, barn-yards. June–August.

D. Tatula L. Roadsides and waste grounds. June–August.

Nicotiana Tabacum Don. Escaped from cultivation. July–August.

SCROPHULARIACEÆ.

Verbascum Thapsus L. Fields and pastures. July.

V. Blattaria L. Waste grounds. June–July.

Linaria Canadensis Dumont. Lookout Mt. June.

L. vulgaris Mill. Fields and fence rows. June–July.

Chelone glabra L. Wet and swampy places. August–September.

C. Lyoni Pursh. Mountain bogs, Alleghanies. August–September.

Scrophularia nodosa L., var. **Marilandica** Gray. Moist thickets. Common. June–July.

Pentstemon lævigatus Solander, var. **Digitalis** Gray. Over the State. June.

P. pubescens Solander. Glades and open rocky places. June–July.

Mimulus ringens L. Along streams and wet places. July–August.

M. alatus Ait. Wet places, edge of ponds. July–August.

Conobea multifida Benth. Sandy, wet ground. Common. July–August.

Herpestis nigrescens Benth. Moist places in the barrens. August–September.

H. rotundifolia Pursh. Swamps along Cumberland river. June–July.

Gratiola Floridana Nutt. East Tenn. and vicinity of Nashville apparently rare. April.

G. Virginiana L. Miry and swampy places. Common. June–July, April–May.

G. ramosa Walt. Barrens at Tullahoma. July.

Ilysanthes gratioloides Benth. Ditches, and along brooklets. April–September.

Veronica Virginica L.   Mountains and highlands.   July.

V. Anagallis L.   Springs and brooklets.   June–July.

V. serpyllifolia L.   Cultivated ground.   April–May.

V. peregrina L.   Fields and pastures.   April–May.

Buchnera Americana L.   Oak barrens.   July.

Seymeria tenuifolia Pursh.   Copses near Cleveland, East Tenn. Sandy soil.   June–July.

S. macrophylla Nutt.   Rich alluvial soil.   Nashville.   July.

Gerardia pedicularia L.   Barrens and Cumberland Mts. June–July.
  Var. **pectinata** Nutt.   Harpeth hills, near Nashville. August.

G. flava L.   Cumberland aud Alleghany Mts., and barrens. June–July.

G. quercifolia Pursh.   Over the State.   July–September.

G. lævigata Raf.   Alleghany Mts.   July.

G. patula Chap.   Rising Sun bluff on Cumberland river, twelve miles below Nashville.   September.

G. purpurea L.   Over the State.   Greenbrier, Robertson Co. July.

G. tenuifolia Vahl.   Harpeth hills.   September.
  Var. **macrophylla** Benth.   Barrens.   July–September.

Castilleia coccinea Spreng.   On siliceous and argillaceous soil, East to West Tenn.   April–May.

Schwalbea Americana Gronov.   Tullahoma.   June.

Pedicularis Canadensis L.   Over the State.   March–April.

Melampyrum Americanum Michx.   Smoky Mts.   July.

OROBANCHACEÆ.

Apyllon uniflorum Gray.   Damp woodlands.   Over the State, but not frequent.   May.

Conopholis Americana Wall.   Oak woods.   Harpeth hills near Nashville.   May.

Epiphegus Virginiana Bart.   On the roots of beech trees. June–July.

LENTIBULARIACEÆ.

Utricularia gibba L. Pond, summit of Lookout Mt. July.

U. biflora Lam. Swamps, West Tenn. July.

BIGNONIACEÆ.

Bignonia capreolata Tourn. Woods, climbing high. April–May.

Tecoma radicans Juss. Woods and cultivated grounds. A noxious weed in fields. May–August.

Catalpa speciosa Ward. Medium size tree. May.

PEDALIACEÆ.

Martynia proboscidea Glox. River banks and waste places. July–August.

ACANTHACEÆ.

Ruellia ciliosa Pursh. Barrens and roadsides. Over the State. July–August.

R. strepens L. Rich soil. June–July.

Dianthera Americana L. Slow-flowing streams. July–August.

Gatesia læte-virens Gray. Lookout Mt. July.

Dicliptera brachiata Sprengl. Rich shady grounds. Frequent in vicinity of Nashville. July–August.

VERBENACEÆ.

Phryma leptostachya L. Moist woodlands. Common. July–August.

Verbena officinalis L. Roadsides and old fields. East Tenn. July.

V. urticæfolia L. Pastures and dry uplands. August–September.

V. angustifolia Michx. Dry soil. Everywhere. June–September.

V. hastata L. Waste grounds. Not frequent. July.

V. stricta Vent. West Tenn. July.

V. bracteata Michx. Around dwellings and along roadsides. June–July.

V. Aubletia L. Very copious in the glades and barrens. May–June.

Lippia lanceolata Michx. Low, moist grounds. August–September.

Callicarpa Americana L. Limestone regions of Middle Tenn. July.

LABIATÆ.

Trichostema dichotomum L. Sandy fields. O. S. July–August.

Isanthus coeruleus Michx. Abundant in Middle Tenn. August–September.

Teucrium Canadense L. Moist meadows and copses. June–September.

Collinsonia Canadensis L. Rich woodlands. July–August.

Mentha viridis L. Wet ground near settlements. July.

M. piperita L. In streamlets. July–September.

M. Canadensis L. Wet places. July–September.

Lycopus Virginicus L. Ponds and ditches. August–September.

L. rubellus Moench. Swampy lands. August–September.

L. sinuatus Ell. Low, swampy grounds. August–September.

Cunila Mariana L. Dry hill, siliceous soil. Abundant in East Tenn. July–September.

Pycnanthemum linifolium Pursh. Oak and cedar barrens, and highlands. June–July.

P. lanceolatum Pursh. Highlands. West Tenn. August–September.

Pycnanthemum muticum Pers. Oak barrens. Very frequent at Tullahoma. August–September.

Var. **pilosum** Gray. Dickson Co., West Tenn. July–September.

P. Tullia Benth. Lookout Mts.; Harpeth hills, south and west of Nashville. August–September.

P. incanum Michx. O. S. July–August.

P. albescens Torr. & Gray? Parksville, East Tenn.; also hill-tops south of Nashville. July–August.

P. montanum Michx. High mountains of East Tenn.; Big Frog Mt.; Clingman-Dom. July.

P. linifolium Pursh. Cedar and oak barrens, and highlands. July–August.

Calamintha Nepeta Link. Dry, rocky ground. Common in vicinity of Nashville. July–September.

C. glabella Benth. Abundant in the cedar glades. May–June.

Hedeoma pulegioides Pers. Dry soil. Common. June–September.

Salvia lyrata L. Copses. Common. April–May.

S. urticæfolia L. River banks and rich soil. May.

Monarda didyma L. Wet places in the high mountains. June–July.

M. punctata L. Vicinity of Memphis. Dr. Egeling.

M. clinopodia L. Mountains of East Tenn. June–July.

M. fistulosa L. Dry copses, fence-rows, etc. Common. August–September.

Var. **mollis** Benth. Hills around Nashville. August–September.

M. Bradburyana Beck. Highlands of Middle Tenn. June.

M. citriodora Cerv. Grass plots, Montgomery-Bell Academy grounds, in Nashville. Adventive. July.

Melissa officinalis L. Escaped from gardens. Nashville. June.

Blephilia ciliata Raf. Dry soil. Very common in Middle Tenn. July.

B. hirsuta Benth. Moist thickets on Cumberland. Rare in our region, Nashville.

Lophanthus nepetoides Benth. Rich soils, thickets and fence-rows, Middle Tenn. August–September.

L. scrophulariæfolius Benth. High mountains of East Tenn. July.

Cedronella cordata Benth. Highlands. July–August.

Nepeta Cataria L. Near dwellings and roadsides. May–June.

N. Glechoma Benth.   Rich, moist thickets.   April–May.

Scutellaria lateriflora L.   Moist woodlands.   July–September.

S. versicolor Nutt.   Rocky places; hills near Nashville.   July–August.

S. saxatilis Riddel.   Chilhowee Mt., East Tenn.   July.

S. serrata Andr.   Woods, Middle Tenn.   July–August.

S. pilosa Michx.   Hills near Nashville, etc.   July–August.
Var. **hirsuta**.   South Tunnel, Sumner Co.

S. integrifolia L.   Highlands, Cumberland Mts.   July–August.

S. galericulata L.   East Tenn., Ducktown.   July–August.

S. canescens Nutt.   Craggy Hope, hills near Nashville.   July–August.

S. parvula Michx.   Very frequent in the cedar glades.   May.
Var. **mollis** Gray.   Dry, rocky places, with the former.

S. nervosa Pursh.   In swampy woodlands, Sumner Co.   July.

Brunella vulgaris L.   Fields and roadsides.   Common.   July–September.

Physostegia Virginica Benth.   Open copses and barrens.   June–August.

Synandra grandiflora Nutt.   Moist woods; John Overton's canebrake, near Nashville.   March–April.

Marrubium vulgare L.   Near dwellings and roadsides.   July.

Perilla ocymoides L.   Escaped from cultivation and spreading.   July.

Leonurus Cardiaca L.   Waste and cultivated grounds.   July–August.

Lamium amplexicaule L.   Common weed in every field or garden.   March–April.

Stachys aspera Michx.   Moist woodlands, etc.   June–August.
Var. **glabra** Gray.   Banks of Cumberland near Nashville.   August–September.

S. cordata Benth.   Moist woodlands and rich hillsides.   (Hills south of Nashville.)   July.

Betonica afficinalis L.   Nashville.   Adventive.   July.

PLANTAGINEÆ.

Plantago cordata Lam. Rare in Tenn. Swampy ground near Brownsville, West Tenn. April–June.

P. major L. Rare. Waste grounds. Adventive. June–August.

P. Rugelii Decaisne. Very frequent around dwellings in the open country. June–August.

P. lanceolata L. Meadows and waste grounds. Introduced. April–June.

P. Patagonica Jacq., var. **aristata** Gray. Along lines of railroad. Tullahoma, Dixon. July.

P. Virginica L. Over the State. Very plentiful in dry sandy soil.

P. pusilla Nutt. Open ground in barrens. April–May. (Mitchellville, Sumner Co.)

P. heterophylla Nutt. Barrens at Lavergne, etc. April–May.

ARISTOLOCHIACEÆ.

Aristolochia Sipho L'Hert. Mountains of East Tenn. Cranberry Iron Works. July.

A. tomentosa Sims. Banks of Cumberland river, near Nashville. May.

A. Serpentaria L. Rich soil in the barrens of Middle Tenn.; also East Tenn. June.

Asarum Canadense L. O. S.

A. arifolium Michx. Damp woods in the mountains of East Tenn. June.

A. Virginicum L. Lookout Mt. Chattanooga.

NYCTAGINEÆ.

Oxybaphus albidus Sweet. Nashville. Bluffs of Mill creek. June.

O. nyctagineus Sweet. Guthrie. July, '83.

PHYTOLACCACEÆ.

Phytolacca decandra L. Everywhere. July–September.

CHENOPODIACEÆ.

Chenopodium murale L. Streets of Nashville. June–September.

*C. **Botrys** L. Around dwellings. Escaped. September.

C. album L. Common. Cultivated ground. September.

C. urbicum L. Streets of Chattanooga. July.

C. ambrosioides L. Brownsville, West Tenn. August.
Var. **anthelminticum** Gray. O. S.

C. glaucum L. . Brownsville, West Tenn. August.

C. Boscianum Moquin. Nashville.

AMARANTHACEÆ.

Amaranthus paniculatus L. Common
Var. **sanguineus** Gray. Cultivated grounds near Nashville.

A. retroflexus L. Fields and gardens.

A. albus L. Streets of Nashville; rocky places. O. S. September.

A. spinosus L. Nashville. O. S. September.

Iresine celosioides L. Rich soils along river banks, Nashville. July.

Montelia tamariscina Gray. Very abundant in cultivated grounds, Nashville. September–October.

POLYGONACEÆ.

Rumex crispus L. Common. June–July.

R. Brittanicus L. Ditches, Nashville. June–July.

R. obtusifolius L. Common; ascends to the highest mountains.

R. verticillatus L. Swamps along Cumberland and Tennessee rivers. July–August.

R. Acetosella L. Pastures. July.

Polygonum orientale L. Escaped from gardens. July.

P. Pennsylvanicum L. O. S. August.

P. incarnatum Ell. River banks. Common near Nashville. July–August.

P. Persicaria L. Waste grounds, near water, O. S. June–July.

P. Hydropiper L. Ditches, etc. Common in streets, etc. August–September.

P. acre H. B. K. Streets of Nashville, etc. July.

P. hydropiperoides Michx. River swamps. In water. Common. August–September.

P. Mühlenbergii Watson. (P. amphibium, var. *terrestre* Gray). River swamps; wild goose pond near Mitchellville. August.

P. Virginianum L. Thickets, rich soil. July–August.

P. aviculare L. Yards and streets everywhere. June–July.

P. erectum L. (P. aviculare, var. *erectum* Roth.) In company with the former, on manured grounds. June–July.

P. tenue Michx. Cumberland plateau, Sewanee. July.

P. arifolium L. Cumberland plateau. September.

P. sagittatum L. Swampy grounds O. S. July–October.

P. Convolvulus L. Low, damp grounds near Nashville. September.

P. dumetorum L., var. *scandens* Gray. O. S. August–September.

Fagopyrum esculentum Moench. Sparingly cultivated and escaped to fence rows. Cumberland Mts. July -August.

Brunnichia cirrhosa Banks. Grounds of Lunatic Asylum, near Nashville. West Tenn. July–October.

LAURACEÆ.

Sassafras officinale Nees. O. S. In the river islands it grows to large dimensions. April.

Lindera Benzoin Meissner. O. S. In rich soil. March–April.

THYMELEACEÆ.

Dirca palustris L. Mountain bogs. Cumberland Mts. April.

SANTALACEÆ.

Comandra umbellata Nutt. In damp soil, oak barrens. Tullahoma. May.

Pyrularia oleifera Gray. Alleghany and Cumberland Mts. Hot Springs, Ducktown.

Buckleya distichophylla Torr. Wolf Creek, Carter Co., East Tenn. Lookout Mt.

LORANTHACEÆ.

Phoradendron flavescens Nutt. O. S. On various deciduous trees. March.

SAURUREÆ.

Saururus cernuus L. O. S. In swamps and ditches. June.

CERATOPHYLLEÆ.

Ceratophyllum demersum L. Swamps along Cumberland river.

CALLITRICHACEÆ.

Callitriche Austini Engelm. Common on mudbanks along river and in moist grounds. Middle Tenn. June.

C. heterophylla Pursh. Pools. Nashville. May.

PODOSTEMACEÆ.

~~Podosteman~~ *abrotanoides* Nutt. In all mountain brooks and streams of East Tenn. July.

EUPHORBIACEÆ.

Euphorbia corollata L. O. S. Argillaceous soils. July.

E. commutata Engelman. Cedar barrens. Middle Tenn. April.

E. dentata Michx. Nashville. Very abundant in the cedar glades. May.

E. humistrata Engelman. River banks and moist pastures. Nashville. July–August.

E. hypericifolia L. Troublesome weed in fields, etc. O. S. July

E. Ipecacuanhœ L. West Tenn., near Johnsonville. May.

E. maculata L. Waste grounds. O. S. July.

E. marginata Pursh. Along railroad near Mitchellville, Robertson Co., Tenn. Abundant about Bowling Green, Ky. July.

E. obtusata Pursh. Paradise ridge, thickets on Charlotte pike, near Nashville. May.

E. serpens H. B. K. Rocky lands, gardens, near Nashville. Abundant. June–September.

E. mercurialina Michx. Vicinity of Nashville, on Stoner's creek, Tunnel hill, also in East Tenn.

E. Lathyris L. Close to Tenn. line in N. C. C. Chickering.

Acalypha Virginica L. O. S. July–August.
Var. **gracilens** Gray. Nashville.

A. Caroliniana Walt. Gardens and fields. O. S. July–August.

Ricinus communis L. Escaped from cultivation.

Stillingia sylvatica L. Vicinity of Memphis. Dr. G. Egeling.

Tragia urticæfolia Michx. Cleveland, East Tenn. July.

T. macrocarpa Willd. Cedar barrens. Middle Tenn. July.

Croton capitatus Michx. Middle and West Tenn. July–September.

C. monanthogynos Michx. O. S. July–August.

C. glandulosus L. Over the State, but not so common as the preceding. July.

Crotonopsis linearis Michx. Cedar barrens of Middle Tenn. Summit of Lookout Mt. July.

Phyllanthus Carolinensis Walter. Pastures and glades. O. S. July–August.

Pachysandra procumbens Michx. White Bluff, Dickson Co.; South Tunnell, Robertson Co.; Beersheba Springs (Col. Wilkin's), Dr. Hampton's farm, Davidson Co. March–June.

URTICACEÆ.

Ulmus fulva Michx. O. S. March.

U. Americana L. O. S. March.

U. racemosa Thomas. Frequent in the vicinity of Nashville. April.

U. alata Michx. Frequent in the glades of Middle Tenn. February–March.

Celtis occidentalis L. O. S. April–May.

Var. Mississippiensis. Much more common and abundant in Middle Tenn. April.

Morus rubra L. Along water-courses. O. S. May.

Maclura aurantiaca Lindl. Frequently planted for hedges. May.

Broussonetia papyrifera Vent. A popular shade tree. Nearly all are male trees, and but few fruit-bearing ones are known in the State. May.

Planera aquatica Gmelin. Brownsville, West Tenn. April.

Urtica gracilis Ait. East Tenn. April. '

U. dioica L. Occasionally introduced, not becoming naturalized.

U. chamædryoides Pursh. Abundant about Nashville. April.

Laportea Canadensis Gaudich. O. S., rich soils. June.

Pilea pumila Gray. O. S., damp thickets. August.

Bœhmeria cylindrica Willd. O. S. with the former. July.

Parietaria Pennsylvanica Mùhl. Waste ground; streets of towns. May–June.

Cannabis sativa L. Escaping from cultivation into hedges, etc. Not observed indigenous.

Humulus Lupulus L. Like the former. Not indigenous. July.

PLATANACEÆ.

Platanus occidentalis L. Largest in bulk of all our timbers. At water's edge on all creeks and rivers. April.

JUGLANDACEÆ.

Juglans cinerea L. O. S. along creek and river banks; nowhere in great numbers. May.

J. nigra L. O. S. Becoming scarce O. S. May.

Carya olivæformis L. Tennessee and Mississippi bottoms in West Tenn. Single trees in Middle Tenn., probably planted by early settlers. Smith's place, Mill creek, near Nashville. May.

C. microcarpa Nutt. Lookout Mt., Chattanooga. May.

C. alba Nutt. Large tree with scaly bark, O. S. April–May.

C. sulcata. Big trees in rich loam O. S.   April–May.

C. tomentosa Nutt. More prevailing than the other, especially in Middle Tenn.   April–May.

C. porcina Nutt.   Barrens and gravelly ridges.   April–May.

C. amara Nutt.   Low grounds, Nashville, along Cumberland river.   April–May.

CUPULIFERÆ.

Quercus alba L. O. S.   April.

Q. stellata Wang. (Q. obtusiloba Michx.) O. S.   April.

Q. lyrata Walt. O. S.   Frequent in vicinity of Nashville. April.

Q. macrocarpa Michx. Low grounds. Nashville. April.

Q. bicolor Willd. Low grounds. Nashville and West Tenn. April.

Q. Michauxii Nutt. O. S.   Rocky hillsides.   April–May.

Q. Prinus L. Hills and mountains of East and Middle Tenn. April.

Q. Mühlenbergii Engelm. Common in Middle Tenn. April.

Q. coccinea Wanger. O. S.   April–May.

Q. tinctoria Bartr. O. S.   April–May.

Q. rubra L. O. S.   April–May.

Q. falcata Michx. O. S.   April.

Q. palustris Du Roi. Low, wet lands. O. S.   May.

Q. ilicifolia Wanger. Scattering. O. S.   April–May.

Q. nigra L. O. S.   April–May.

Q. imbricaria Michx. Barrens and glades. April–May.

Q. Phellos L. Low, wet ground in the barrens.   April–May.

Q. aquatica Castesby. Mountains of East Tenn.; along creeks and barrens of Middle Tenn.   April–May.

Our oaks are often difficult to determine. Many transitions or hybrid forms occur.

Castanea vulgaris Lam., var. Americana DC. O. S.   June.

C. pumila Michx. Common in East Tenn. June.

Fagus ferruginea Ait. O. S.   May.

Carpinus Americana Michx. From highest mountain summits to the lowlands. April.

Ostrya Virginica Willd. Limestone glades. April.

Corylus Americana Walt. Rich soils. O. S. April.

C. rostrata Ait. Common in the mountains of East Tenn. April.

BETULACEÆ.

Betula lenta L. Lookout Mt., Alleghanies.

B. lutea Michx., fil. Summit of Smoky Mts. July.

B. nigra L. O. S. Abundant in West Tenn. April.

B. papyracea Ait. O. S. Principally mountains of East Tenn. May.

Alnus viridis DC. Mountains of East Tenn.

A. serrulata Ait. O. S. March.

SALICACEÆ.

Salix nigra Marsh. River banks. The form *S. nigra*, var. *Wardi* Bebb., in island of Cumberland at Nashville. April–May.

S. alba L. O. S. April.

S. humilis Marsh. In the barrens and high mountains. May.

S. longifolia Mühlb. West Tenn.

S. tristis Ait. Barrens of Middle Tenn. March.

S. lucida Mühlb. Mountains of East Tenn.

S. petiolaris Smith. Buena Vista ferry road, Nashville. April.

S. purpurea L. Introduced and cultivated for basket work.

S. Babylonica L. Introduced.

Populus monilifera Ait. O. S. March.

P. heterophylla L. O. S. March–April.

P. balsamifera L., var. *candicans* Gray. Introduced by early settlers in East Tenn. (Balm of Gilead.)

P. alba L. Introduced and spreading. March.

P. dilatata Ait. Introduced. Short-lived in Middle Tenn.

6

ARACEÆ.

Arisæma triphyllum Torr. O. S. April–May.

A. polymorphum Chapm. Roane Mt. 4–5000 ft. June (in fruit). Legit. J. W. Chickering, Jr.

A. Dracontium Schott. Nashville, East Tenn. May.

Peltandra Virginica Raf. Cleveland, East Tenn., Capt. Raht's spring branch, and Robertson Co. July.

Orontium aquaticum L. Cleveland, East Tenn. With the former.

Acorus Calamus L. In an old garden in Nashville, Market street. Said to grow abundantly in the vicinity of Red river, Montgomery Co.

LEMNACEÆ.

Lemna perpusilla Torr. Shelby pond, Nashville. O. S.

L. trisulca L. Ponds, West Tenn. Submerged.

L. minor L. Ponds O. S.

Speirodela polyrrhiza Schleid. O. S.

Wolffia Columbiana Karst. Submerged. Pond on Lebanon pike, near Nashville. September.

W. Brasiliensis Weddel. Floating. Pond in Lunatic Asylum grounds covered with it. September, 1884.

TYPHACEÆ.

Typha angustifolia L. Near Lavergne, Williamson Co.; Ducktown, East Tenn. Very rare.

T. latifolia L. O. S.

Sparganium simplex Huds., var. *Nuttalli* Gray. Along Cumberland river, Cleveland, East Tenn. July.

S. eurycarpum Englm. East Tenn , Tullahoma creek. July.

Echinodorus radicans Englm. Swamps on Cumberland river, above water-works at Nashville. Common. July.

ALISMACEÆ.

Sagittaria variabilis Englm. Ponds and swamps. Common. August–September.

Var. *angustifolia* Englm. Hollow Rock. July–September.

S. heterophylla Pursh.   With the former.   July–September.

S. graminea Michx.   With the former.   August–September.

Alisma Plantago L.   Var. *Americana* Gray.   Common.

HYDROCHARIDEÆ.

Vallisneria spiralis L.   Slow-flowing streams.   East Tenn. Infrequent.   July.

Anacharis Canadensis Planchon.   Pools above water-works, Nashville, East Tenn., O. S.   June.

NAIADACEÆ.

Potamogeton pauciflorus Pursh.   Nashville, also East Tenn. Frequent.   July.

P. Claytonii Tuckerm.   Mountain streams of East Tenn.   July.

P. pusillus L.   Wolf Creek, East Tenn.   July.

P. hybridus Michx.   Tullahoma.   July.

P. natans L.   Wolf river, near Memphis.   Dr. Egeling.

P. perfoliatus L.   Wolf river, vicinity of Memphis.   Dr. Egeling.

Zannichelia palustris L.   In ponds and springs about Nashville.   April.

ORCHIDACEÆ.

Orchis spectabilis L.   River bottoms below Nashville.   Very rare.   April.

Habenaria integra Sprengl.   Tullahoma ; Mitchellville.   July.

H. virescens Sprengel.   Swamps.   West Tenn.   July.

H. cristata R. Br.   Sewanee.   Cumberland Mts.   July–August.

H. peramœna Gray.   Cedar Hill, Mitchellville, Robertson Co. June.

H. ciliaris R. Brown.   Edgefield Junction.   O. S.   June–July.

H. blephariglottis Hooker.   Tullahoma.   August.

H. tridentata Hooker.   Sewanee ; Cumberland Mts.   Common.   July.

Goodyera pubescens R. Br.   Mountains of East Tenn.   Wolf creek, etc.   July.

Spiranthes simplex Gray. Hills south of Nashville. Frequent. August.

S. cernua Rich. Tullahoma. Barrens. Common. October.

S. graminea Lindl. Tullahoma. Barrens. September.

S. gracilis Bigl. Mount Olivet Cemetery. Barrens. September.

Pogonia verticillata Nutt. Sewanee leg. Gen. Kirby-Smith. May.

⊶ *P. ophiaglossoides* Nutt. Ducktown. June.

P. pendula Lindley. Mitchellville. Wartrace. September.

P. divaricata R. Br. Mountains of East Tenn. June.

Calopogon pulchellus R. Br. Tullahoma; Parksville. Frequent. July.

Tipularia discolor Nutt. Ducktown, Polk Co. August.

Bletia aphylla Ruiz & Pavon. Parksville; also in the "Flatwoods," Bradley Co. Always under pines! July.

Corallorhiza innata R. Br. Hills south of Nashville; Mitchellville. September.

C. multiflora Nutt. O. S. June.

C. odontorhiza Nutt. Hills near Nashville. April.

Cypripedium acaule Ait. East Tenn. May.

C. spectabile Swartz. Ducktown. July.

C. pubescens Willd. Rich woods in the mountains. (Lookout Mt.) May.

C. parviflorum Cones. Hills near Nashville. May.

AMARYLLIDACEÆ.

Pancratium rotatum Ker. Low, wet meadow lands. O. S. June.

Agave Virginica L. Dry, rocky places. O. S. July–August.

Hypoxis erecta L. Frequent in cedar and oak barrens. May.

HÆMODORACEÆ.

Aletris farinosa L. Over the State, especially oak barrens. May–June.

IRIDACEÆ.

Iris versicolor L. Swamps. Tullahoma, Mitchellville, etc. May.

I. cuprea Pursh. Swamps. West Tenn. June.

I. cristata Ait. O. S. May.

I. Virginica L. Tullahoma. Low, wet ground. Abundant. June–July.

I. Germanica L. Deserted homestead, Charlotte pike, near Nashville. April.

I. hexagona Walt. Swamp east side of Tennessee river at Johnsonville. July.

Pardanthus Chinensis Kerr. Over the State. Indigenous. July.

Sisyrinchium anceps L. Barrens of Middle Tenn. May–June.

S. mucronatum Michx. East Tenn. May–June.

DIOSCOREACEÆ.

Dioscorea villosa L. Rich woodlands. O. S. April.

SMILACEÆ.

Smilax rotundifolia L. O. S. May.
Var. **quadrangularis** Mühlb. Nashville. September.

S. glauca Walter. O. S. Creeping low over fields. May.

S. Pseudo-China L. River banks. May.

S. hispida Mühlb. Low, damp ground.

S. herbacea L. Nashville, in rich, moist ground.

S. tamnifolia Michx. Nashville. In rich woodlands.

LILIACEÆ.

Allium cernuum Roth. Over the State. July.

A. Canadense Kalm. Common. June.

A. tricoccum Ait. Rare, apparently. East Tenn., Ducktown. July.

A. mutabile Michx. Cedar glades, Lavergne. May

A. sativum L. Introduced. Grounds of Mrs. Cheatham, and old cemetery, Nashville. June.

Nothoscor diumstriatum Kunth. Moist ground; frequent in Middle Tenn. May.

Camassia Fraseri Torr. Rich woodlands here and there over the State. April.

Schœnolirium croceum Gray. Moist places in the cedar glades, Lavergne. May.

Convallaria majalis L. Frog Mts., East Tenn., near Ducktown; also said to occur at Sewanee. May.

Polygonatum giganteum Dict. Rich thickets. Common. May.

P. biflorum Elliott. With the former. April.

Smilacina racemosa Desf. Rich woodlands over the State. May–June.

Maianthemum Canadense Desf. Summit of Big Thunderhead, Smoky Mts. June.

Asparagus officinalis L. Escaped from cultivation. June.

Yucca filamentosa L. Dry, rocky ground O. S. Very frequent in Middle Tenn. May.

Lilium Grayi Watson. Summit of Roane Mt. Dr. Gray, Prof. Chickering. July.

L. superbum L. In splendid and very numerous specimens on Big Frog Mts., East Tenn. July.
Var. Carolinianum. Throughout the mountains. July.

L. Canadense L. Paradise ridge, near Nashville. July.

Erythronium Americanum Smith. O. S. Harpeth hills, near Nashville. April.

E. albidum Nutt. White's Bend, below Nashville; summit of Roane Mt., East Tenn. Carby. April.

Ornithogalum umbellatum L. Buchanan's fort on Mill creek, Nashville. Escaped from cultivation. April.

Uvularia perfoliata L. South Tunnell, Sumner Co. May.

U. grandiflora Smith. Rocky banks of Cumberland, below Hydt's ferry. May.

Oakesia sessilifolia S. Watson. Oakland Station, Robertson Co. May.

Streptopus roseus Michx.   High points of the Smoky Mts.; Big Thunderhead.   June.

Prosartes lanuginosa Don.   Nashville, Sewanee and mountains of East Tenn.   May.

P. maculata Gray.   Mountains of East Tenn.; *vide* Am. Jour. Sc. 2, 45.

Clintonia umbellata Torr.   Big Frog Mt.; Smoky Mts.   June.

Medeola Virginica L.   Mountains of East Tenn.; Cumberland Mts.   May.

Trillium cernuum L.   Lookout Mt.; Ducktown.   May.

T. sessile L.   Over the State.   April.   ·
Var. **Wrayi** S. Watson.   Hills near Nashville.   April.

T. erectum L.   Vicinity of Nashville.   Rare.

T. stylosum Nutt.   Memphis.   Dr. Egeling.

T. erythrocarpon Michx.   Ocoe Valley.   May.

Melanthium Virginicum L.   Big Frog Mt.   July.

M. parviflorum Gray.   With the former.   July.

Veratrum viride Ait.   Wolf creek on Bench Mt.   June.

Stenanthium angustifolium Gray.   Chilhowee Mts.   July.

S. robustum Watson.   Wet ground in the barrens, Tullahoma; South Tunnell, Sumner Co.; also mountains of East Tenn.   July.

Amianthium muscætoxicum Gray.   Bogs.   Common in East Tenn.   May–June.

Zygadenus angustifolius Watson.   Barrens at Tullahoma.   June.

Chæmælirium Carolinianum Willd.   Dry woodlands.   Over the State.   July–August.

Hemerocallis fulva L.   Escaped.

H. flava L.   Escaped.

Muscari botryoides Mill.   Escaped from gardens into fence-rows.

PONTEDERACEÆ.

Pontederia cordata L.   Swamps.   East Tenn.   Wild-goose pond near Mitchellville, Robertson Co.

Heteranthera reniformis Ruiz & Pav.   Swamps and ditches. O. S.   Nashville; along Cumberland river.   August.

H. limosa Vahl. With the former in Middle and West Tenn.

Schollera graminea Willd. East and West Tenn. Apparently rare in Middle Tenn.

COMMELYNACEÆ.

Commelyna Cayennensis. Rich wet grounds. Common. Nashville, up and down the river. July.

C. erecta L. Shaded river banks. Nashville. August–September.

C. Virginica L. Copses. O. S. June–July.
Var. **angustifolia** Michx. Cedar barrens. June–July.

Tradescantia Virginica L. O. S. Rich woodlands. May.

T. pilosa Lehm. Rich, shady soil. O. S. July–September.

XYRIDACEÆ.

Xyris Caroliniana Walt. Mountain meadows and brooks of East Tenn. July.

X. flexuosa Mühlb. Swamps in Hadley's bend, near Nashville; low ground in the barrens. August–September.

JUNCACEÆ.

Luzula campestris DC. O. S. April–May.

L. pilosa Willd. Cumberland and Alleghany Mts. May–June.

Juncus effusus L. Ponds and ditches, vicinity of Nashville. June.

J. setaceus Rostkov. Barrens and mountains of East Tenn. June–July.

J. tenuis Willd. Roadsides in damp soil. June.
Var. **secundus** Engelm. Oakland Station, Roberston Co. June.

J. dichotomus Ell. O. S. Frequent around Nashville. June.

J. scirpoides L., var. *macrostemon* Engelm. Cleveland, East Tenn. July.

J. buffonius L. East Tenn. May.

J. marginatus Rostk. Mountains of East Tenn. and oak barrens. June.

J. acuminatus Michx., and varieties O. S. June–July.

J. repens Michx. Swamps O. S. June–July.

J. pelocarpus E. Meyer. Tullahoma. July.

J. leptocaulis Torr. & Gray. Frequent in the cedar glades, with Isotes Buttleri and Leavenworthias. May–June.

J. brachycarpus Engelm. Ponds along Cumberland river. June–July.

J. articulatus L. Cleveland, East Tenn. July.

J. Canadensis J. Gay. East Tenn. September.

CYPERACEÆ.

Cyperus flavescens L. Ponds and ditches. July–September.

C. diandrus Torr. With the former. August–September.
Var. **casteneus** Torr. With the preceding. September.

C. aristatus Rottb. (*C. inflexus* Mühlb.). Glades and river banks. July.

C. Schweinitzii Torr. East Tenn. July–September.

C. Luzulæ Rottb., var. *umbellulatus* N. L. Britton (*C. vegetus* Pursh.). Damp argillaceous soils. July.

C. virens Michx. West Tenn. Rare. August.

C. acuminatus Torr. & Hooker. Cedar glades. Lavergne. July.

C. rotundus L. Low grounds near Nashville; Horticultural garden. Not frequent. July.

C. esculentus L. (*C. phymatodes* Mühl.). Intrusive weed. July–September.
Var. **angustispicatus** N. L. Britton. Swamps along Cumberland river. July–September.

C. strigosus L. (*C. Michauxianus* Schult.). Common in low, damp ground.
Var. **robustior** Kunth. River swamps.
Var. **capitatus** Boekl. Sandy river banks.
Var. **compositus** N. L. Britton. With the above.
Var. **elongatus** N. L. Britton (*C. Michauxianus*, var. *elongatus* Torr.). Banks and islands in Cumberland river.
Var. **pygmæus** n. var. Only 3–4 inches high, resembling *C. Schweinitzii*. McSpadden's bend, near Nashville.

C. refractus Engelman. Nashville, July–September. Hollow Rock, West Tenn., August–September.

C. erythrorhizos Mühl. River swamps; 1–4 feet high. August-September.

C. speciosus Vahl. (*C. Michauxianus* Torr.). Sandy banks and river swamps. Nashville. September.

C. ovularis Torr., var. robustus, Backl. Moist spots in the cedar glades.

Var. sphæricus Backl. With the above.

C. filiculmis Vohl. Dry uplands. June–July.

C. Lancastriensis Port. Dry, rocky and waste grounds, vicinity of Nashville. July.

Killingia pumila Michx. Miry places. June–September.

Dulichium spathaceum Pers. Deep river swamps. June–September.

Hemicarpha subsquarrosa Nees. Davidson Co.

Elæocharis obtusa Schult. Bogs and wet ground. May–July.

E. Engelmanni Steudel. Damp places in the barrens. June–July.

E. palustris R. Br. Common. June–September.

E. tenuis Schultes. Barrens and highlands. July.

E. acicularis R. Br. Low, wet places. July–October.

E. quadrangularis R. Br. River swamps. July–September.

E. intermedia Schult. Frequent. July–August.

E. compressa Sullivant. Springy places in calcareous soil. May–June.

Rhynchospora corniculata Gray. Swamps. August–September.

R. cymosa Nutt. With the above. August–September.

R. fusca Roem. & Schult. Mountain bogs. Cumberland Mts. July.

R. glomerata Vohl. Bon Air, Tullahoma; Lookout Mt. August.

R. alba Vohl. Mountain bogs, Cumberland and Alleghany Mts. July.

Scleria triglomerata Vohl. Lookout Mt.; Tullahoma. July.

S. pauciflora Mühlb.   Lavergne.   May–June.

Scirpus validus Vohl.   Cleveland, Capt. Raht's place.   July.

S. debilis Pursh.   Swamps West Tenn.   August–September.

S. atrovirens Mühlb.   Frequent.   July–August.

S. fluviatilis Gray.   Ducktown, East Tenn.

S. polyphillus Vohl.   Dickson Station.   Very common in the mountains.   June–July.

S. lineatus Michx.   Borders of springs in the glades.   June–July.

S cæspitosus L.   Roane Mt., East Tenn.   Chickering.

Eriophorum Virginicum L.   Cumberland Mts.   July.

E. polystachyum L.   Ducktown, East Tenn.   July.

Fimbristylis capillaris Gray.   Sandy places in the mountains.   June–July.

F. autumnalis Roem. & Schult.   Bogs and ditches.   September.

F. laxa Vohl.   Cedar glades, Lavergne.   July–August.

Dichromena latifolia Baldw.   Oak barrens, Middle Tenn.; Tullahoma.   Not frequent.   August.

Carex polytrichoides Mühlb.   Dry hills near Nashville.   June.

C. Steudelii Kunth.   Highlands.

C. teretiuscula Goodw.   Mountains of East Tenn.

C. vulpinoidea Michx.   Dry copses, Charlotte pike.   Nashville.   June.

C. cephalophora Mühlb.   Nashville, Kingston Springs.   May.

C. cephaloidea Boot.   Around Nashville, Paradise ridge.   June.

C. rosea Schk.   East Tenn.   July.

C. Mühlenbergii Schk.   Nashville.   June–July.

C. retroflexa Mühlb.   Nashville.   June.

C. trisperma Dew.   High mountains of East Tenn., Frog Mt., Big Thunderhead; 6000'.   July.

C. straminea Schk.   Nashville, Charlotte pike.   May.

C. crinita Lam.   Cumberland Mts., Whiteside.   July.

C. **Shortiana** Dew. Tunnell Hill, Sumner Co., East Tenn. June.

C. **Crawei** Dew. Dry barrens of Middle Tenn., Lavergne Station. June–July.

C. **granularis** Mühlb. Moist meadows, highlands, and in East Tenn. June–July.

C. **grisea** Wahlb. Low grounds adjoining Cumberland river near Nashville. June.

C. **flaccosperma** Dew. Moist thickets, vicinity of Nashville; on the ridge. June.

C. **Davisii** Swaegr. Jones' Bend, near Edgefield Junction, edge of a swamp. June.

C. **gracillima** Swaegr. Jones' Bend, border of swamps. June.

C. *œstivalis* M. A. Curtis. Roane Mt. Prof. Chickering. Clingman-Dom. 6000'. July.

C. *virescens* Mühlb. Cumberland Mts., valley of East Tenn. June.

C. **triceps** Michx. Dry glades of Middle Tenn., Lavergne. June.

C. *plantaginea* Lam. Mountains near Ducktown, East Tenn. July.

C. **laxiflora** Lam. Low, wet woodlands and copses around Nashville; highlands of Robertson Co. June.

C. **digitalis** Willd. Low grounds throughout the State. July.

C. **oligocarpa** Schk. Swampy places on the ridge, Sumner Co.; Jones' bend, East Tenn. June.

C. **Emmonsii** Dew. Dry hills near Nashville, in siliceous soil. June–July.

C. **nigro-marginata** Schwarz. Dry copses, vicinity of Nashville. May–June.

C. *miliacea* Mühlb. Ducktown, East Tenn.; perhaps throughout. June.

C. *juncea* Willd. Roane Mt. Prof. Chickering. August.

C. **debilis** Michx. Border of swamps near Mitchellville, Sumner Co. June–July.

C. **tentaculata** Mühlb. Throughout the State. South Tunnel, Edgefield; Shelby pond. June.

C. **intumescens** Rudge. Swamps along Cumberland river; Shelby pond. June.

C. *Grayi* Carey. Swamps of West Tenn. (Humbolt). July.

C. **lupulina** Mühlb. Swamps along Cumberland and on highlands; Shelby pond. May–June.

C. **stenolepis** Torr. River swamps throughout the State. June–July.

C. **squarrosa** L. With the former. June–July.

C. **bullata** Schk. River swamps; Jones' bend. June–July.

GRAMINEÆ.

Paspalum fluitans Kunth. Slow streams. July–September.

P. **distichum** Linn. Low grounds; margin of ponds. September.

P. **setaceum** Michx. (P. debile Michx.) Damp, sandy soil in the upland barrens. September.

P. **ciliatifolium** Mühlb. Very common. July–September.

P. **Walterianum** Schult. Low, wet grounds. September.

P. **læve** Michx. Damp ground. Very variable. August–September.

Var. **undulosum.** Oak barrens. Dry situation. (*P. undulosum* Le Conte.)

Var. **angustifolium.** In wet lands. (*P. angustifolium* Le Conte.)

Var. *pilosum* Vasey. Sheets covered with long, soft hairs. Barrens.

Var. *radicans* Vasey. Decumbent, branching at the lower nodes; culms erect, $2\frac{1}{2}$–3 feet high; leaves purplish. Swamps at Hollow Rock. September.

P. **dilatatum** Poir. Open grounds and grass plots. August–October.

Panicum agrostoides Spreng. Inundated on wet land. Variable. August–November.

P. **anceps** L. Damp soils. Cedar glades, etc. Lavergne. August–October.

P. **capillare** L. Exceedingly abundant. Preferring dry uplands and barrens. I distinguish four distinct varieties.

Var. campestre. Root leaves forming flat tufts; appressed to the soil, forming tufts. This is the most common form.

Var. agreste. Stout and very hairy; panicle very large and widely divaricated; forming no tufts. Common. In loose ground, fields and gardens.

* Var. flexile Gattinger. Thin, elastic and upright, with smaller panicle and acute flowers. Closely resembles P. autumnale Bosc. Characteristic of the cedar glades. July–September.

Var. minimum Engelman. Dwarf, resembling P. depauperatum. Poor, loamy and siliceous soil of the highlands. July–August.

P. clandestinum L. Along the banks of Cumberland river, in the canebrakes; also in East Tenn., in rich bottom lands. July–September.

Var. pedunculatum Torr. With the former.

P. colonum L. Ponds and ditches around Nashville, Lavergne, etc. September–October.

P. commutatum Schultz. (*P. nervosum* Mühlb.) Woods, rich and loose soil, over the State. May–August.

P. crus-galli L. Ponds and ditches everywhere. July–September.

Var. hispidum Mühlb.
Var. muticum Vasey. With the former.

P. depauperatum Mühlb. Dry copses and woodlands. Highlands. May–June.

P. dichotomum L. Over the State, in many varieties.
Var. nitidum Lam. In the cedar glades.
Var. pubescens. Woods, etc.
Var. barbulatum Gray. Highlands, in siliceous soil.
Var. sphærocarpon Gray. Earliest in the glades. April–May.

P. filiforme L. Argillaceous and siliceous soils. August–September.

---

* Panicum capillare L., var. flexile Gattinger, is either one of the forms of P. capillare, or an annual variety of P. autumnale Bascoe, which it resembles greatly, especially in smoothness and form of spikelets. It is very smooth above, with some hairs on the lower part of culm and leaves; culm very slender, panicle rather small, and branches not spreading until the mature spikelets are ready to drop off; leaves linear, gradually attenuate, of a pale green color. It abounds in the cedar glades, and is rarely seen outside of them.

P. glabrum Gaudin., var. *Mississippiense* Gattinger. Confined to strong argillaceous soils, and immediately disappearing where calcareous soils commence. Vicinity of Nashville, etc. September–October.

P. latifolium L. Thickets, edge of woodlands, appearing early. May–June.

P. laxiflorum Lam. Damp, rich woodlands; highlands. July–August.

P. microcarpon Mühlb. (*P. multiflorum* Ell.). Rich and moist localities. Not frequent. Ocoe valley, East Tenn.; Craggy Hope, Cheatham Co. July–August. Not to be mistaken for **P. dichotomaerum**, var. microcarpon, which appears early in April–May.

P. proliferum Lam. Wet lands. Very common. August–October.

Var. **geniculatum** Ell. With the former.

P. sanguinale L. Cornfields and roadsides everywhere. September.

P. scoparium Mühlb. (*P. pauciflorum* Ell.). Cedar barrens near Lavergne. August–September.

P. verrucosum Mühlb. Swampy lands along Cumberland river, Jones' Bend, Mitchellville. September.

P. virgatum L. Moist, sandy soil along Cumberland river; highlands and barrens at Tullahoma. July–August.

Setaria glauca Beauv. Common. Fields and waste places. August–September.

Var. **lævigata** Chapm. Glades of Middle Tenn. More common than the former.

S. viridis Beans. Fields and roadsides. Frequent. July–September.

S. verticillata Beans. Cultivated lands. Not frequent.

S. Italica Kunth. Frequently cultivated and escaped. August–September.

Penicillaria spicata Kunth. Cultivated and sometimes escaped. July–August.

Cenchrus tribuloides L. Sandy banks of Mississippi river. July–August.

Spartina cynosuroides Willd. Brownsville, West Tenn. September–October.

Tripsacum dactyloides L. West Tenn., near Hickman ; also, Nashville, in old graveyard. Large patches. July–August.

Leersia oryzoides Swartz. Swamps and banks of creeks. Very common. August–September.

L. Virginica Willd. Damp, low ground, in the shade, everywhere. August–September.

Erianthus alopecuroides Ell. In places over the State ; Oakland Station, Tullahoma, in siliceous soil. August–September.

E. brevibarbis Michx. Wild-goose pond near Mitchellville, Sumner Co.; only locality known to me in Middle Tenn. In company with *Arundinaria tecta*. September.

E. strictus Baldwin. Old fields near Tullahoma, Cleveland, East Tenn. August–September.

Danthonia compressa Aust. Higher mountains of East Tenn. July.

D. spicata Beauv. Harpeth hills in poor siliceous soil. June.

D. sericea Nutt. In a cedar glade beyond Edgefield Junction, Davidson Co. June–July.

Cynodon Dactylon Pers. Sandy banks of the Mississippi river; around dwellings over the State. July.

Chloris verticillata Nutt. Garden of Mr. Rath, in Cleveland, East Tenn. Introduced. July.

Gymnopogon racemosus Beauv. Barrens east of Tullahoma. July.

Bouteloua racemosa Lag. (*B. curtipendula* Gray). Cedar glades. June–July.

Eleusine Indica Gaert. Very common in fields and gardens as a common weed. July–August.

Leptochloa mucronata Kunth. Common in the cornfields and cultivated grounds of Middle Tenn. July–September.

Triodia seslerioides Benth. Over the State in all soils. September–October.

Eatonia Pennsylvanica Gray. Copses around Nashville, etc. May.

E. Dudleyi Vasey. Highlands. April–May.

E. obtusata Gray, var. *laxiflora* Gattinger. Highlands. Very rare. May.

Eragrostis reptans Nees.   Wet, sandy soil on river banks. September–October.

E. poæoides Beauv.   Cultivated ground.   Common.   August–September.   (Eragrostis Brownei Nees is indistinguishable from poæoides, and occurs in the cedar glades.   Very common.)
Var. **megastachya**.   Cultivated ground.   With the former.

E. Frankii Meyer.   Exceedingly copious in dry lands and glades of Middle Tenn.   July–August.

E. pectinacea, var. *refracta* Chapm.   Common in the cedar glades.   July–August.

E. Purshii Schrad.   Streets of Nashville.   In dry and wet soil.   Exceedingly frequent.   August–September.

E. tenuis Gray.   Harpeth Hills and highlands.   July–August.

E. oxylepis Torr.   Vicinity of Memphis.   Dr. G. Egeling.

Andropogon claudestinus Hale (*Androp. Elliottii* Chap.). Barrens at Tullahoma.   October.

A. dissitiflorus Michx. (*A. Virginicus* L.).   Common.   September–October.
Var. **vaginatus** Chap.   With the former.

A. macronrus Michx.   Sandy old fields.   Over the State. September–October.

A. provincialis Lam.   Edge of roads and fence rows.   Common.   August–October.

A. scoparius Michx.   Old fields.   September–October.
Var. **multiramea** Hack.   Banks of Cumberland.   September–October.

Sorghum Halepense L.   Naturalized in vicinity of Nashville. August–September.

Chrysopogon avenaceus Benth.   Open barrens.   Over the State.   July–September.

Phalaris Canariensis L.   Near dwellings.   Introduced.   August–September.

Ph. arundinacea L.   The garden variety, Ph. arundinacea picta, sometimes found escaped.

Anthoxanthum odoratum L.   In meadows, in East Tenn. Naturalized.   July.

7

Alopecurus pratensis L. Rare. Sometimes introduced. Flowers early. May.

A. geniculatus, var. aristulatus Michx. Ponds and ditches. May.

Aristida dichotoma L. Common in poor soils. September–October.

A. gracilis Elliott. With the former, in argillaceous soils. September–October.

A. ramosissima Engelm. Humboldt, West Tenn. September.

A. purpurascens Poir. Paradise Ridge, Robertson Co. August–September.

Stipa avenacea L. Valley of East Tenn.; Lookout Mt. June.

Mühlenbergia capillaris Kunth. Cedar glades, Lavergne. September–October.

M. diffusa Schreber. Grass plots, pastures, everywhere. September–October.

M. Mexicana Trin. Thickets along river banks, etc. Common throughout. September.

Mühlenbergia sobolifera Trin. Rocky woodlands. Harpeth hills, etc. September–October.

M. sylvatica Torr. & Gray. Damp woodlands; river bottoms. September–October.

M.Willdenovii Trin. With the former, and ascending 6000′ in the Frog and Smoky Mts. July–September.

Brachyelytrum aristatum Beauv. Dry woodlands, in places. August–September.

Phleum pratense L. Naturalized and frequently cultivated. July.

Sporobolus Indicus R. Brown. Sandy soil in the Cumberland Mts. and in the oak barrens. July–September.

S. vaginæflorus Torr. Dry pastures, Middle Tenn.
Var. **exsertus.** Cedar glades. September–October.

Agrostis arachnoides Ell. Argillaceous soil, in the glades and highlands. Kingston Springs; Lavergne. May–June.

A. canina, var. *rupestris* Chapm. High mountains of East Tenn.; Roane Mt. Prof. Chickering. July.

A. perennans Tuck. Open woodlands. Over the State. July–October.

A. scabra Willd. Sandy soils, here and there. June–July.

A. vulgaris With. Largely cultivated and indigenous. July. Var. **alba** L. Dry woodlands. Over the State. July–September.

Cinna arundinacea L. Low grounds and bottoms. Variable. A slender form approaching C. pendula occurs on Paradise Ridge. August–September.

Deyeuxia Nuttalliana Vasey. Hills and mountains of East Tenn. July–September.

Deschampsia flexuosa Beauv. Mountains of East Tenn. July–September.

Holcus lanatus L. Extensively naturalized and spreading in East Tenn. June–July.

Trisetum palustre L. Mountains of East Tenn. Overhanging wet rocks on Ocoe river. July.

Avena sativa L. Extensively cultivated, sometimes escaped. June.

Arrhenatherum avenaceum Beauv. Clifton pike, beyond Jubilee Hall, Nashville. June.

Melica mutica Walt. Over the State, shady hill-sides and ravines. April–May.

Diarrhena Americana Beauv. Rich soil amongst rocks, here and there. September–October.

Uniola latifolia Michx. Cliffs on Cumberland river and Mill creek. Generally over the State in like localities. June–July.

U. gracilis Michx. Damp soil in the oak barrens. July–August.

Dactylis glomerata L. Generally cultivated and dispersed from cultivation. Don't seem to be inclined to become thoroughly naturalized like Holcus lanatus.

Poa annua L. Common in all waste grounds and roadsides. April–May.

P. compressa L. Dry meadow lands. Naturalized and spreading. June–July.

P. flexuosa Mühlb. Damp ground, edge of ponds, especially in the highlands. June–July.

P. pratensis L. Over the State. Cultivated and indigenous. May–June.

P. sylvestris Gray. Abundant in all woodlands and copses. May.

P. trivialis L. Very easily mistaken for pratensis, and therefore frequently overlooked. Cockrill's farm, near Nashville. May–June.

Glyceria acutiflora Torr. In a pond, near the water-works at Nashville. June.

G. aquatica Sm. var. *Americana* Vasey, (*Glyceria arundinacea* Kunth.). In a small branchlet near Cumberland river, Bell's bend. September.

G. nervata Trin. Common over the State in wet meadows. July.

G. pallida Trin. In a mountain bog near Ducktown, East Tenn.

Festuca Myurus L. Near Lunatic Asylum, Nashville.

F. tenella Willd. Poor argillaceous soils; hill-tops south of Nashville. April.
Var. **aristulata** Torr. With the former.

F. elatior L. In meadow lands. Introduced and naturalized. June.

F. Shortii Vasey. Barrens at Tullahoma. First collected in 1867, but not recognized, and mistaken for F. elatior, resembling it in general habits. July.

F. ovina L. In the barrens and on cliffs on the Cumberland river. June.

Bromus ciliatus L. Woodlands over the State. May–June.
Var **purgans** Gray. With the former.

B. mollis L. Cultivated grounds. Not frequent. May.

B. secalinus L. Abundant in grain fields. June.

B. racemosus L. Fields and pastures. Not so abundant. June.

B. sterilis L. Introduced on the grounds of Mr. Washingington, Granny White Pike, Nashville. June.

Lolium perenne L. Introduced and perhaps spreading. Capitol grounds. May–June.

L. temulentum L.   From the farm of Mr. Lenoir, near Knoxville.   July.

Agropyrum caninum R. & S.   In cultivated ground.   Not frequent.   June.

A. repens Beauv.   Cultivated grounds.   Like the former, not very frequent.   June.

Hordeum pratense Huds.   Thin lands.   Charlotte pike, near toll-gate.   Common.   May–June.

Elymus Canadensis L.   Common over the State.   July.

Var. **glaucifolius** Gray.   Cedar glades.

E. striatus Willd.   Over the State.   July.

Var. **villosus.**   Dry rocky places, with the former.

E. Virginicus L.   Abundant in Middle Tenn., loving strong limestone soil.   June–July.

Asprella Hystrix Willd.   Rocky glens over the State.   June.

Arundinaria macrosperma Michx.   Along the large streams of the State.   April.

A. tecta Mühlb.   Lookout Mt.   Wild goose pond near Mitchellville, Sumner Co.   Readily overlooked where it grows intermixed with other vegetation.   It grows in the water.

GYMNOSPERMÆ.

CONIFERÆ.

Juniperus Virginiana L.   Red Cedar.   Scattered over the State, and forming extensive cedar forests in Middle Tenn.   Fl. April.

Pinus rigida Miller.   Pitch Pine.   Cumberland Mts. and mountains of East Tenn.   On the spurs of Big Thunderhead, in Carter, it accompanies the following:

P. pungens Michx.   Table-Mountain Pine.   Big Thunderhead.   Abundant.

P. inops Ait.   Jersey or Scrub Pine.   Over the State, especially on sterile rocky mountain lands.   April–May.

P. mitis Michx.   Yellow Pine.   Cumberland Mts. and East Tenn.   April–May.

P. Strobus L. White Pine. Cumberland Mts., and especially the shady and moist gorges of the high mountains of East Tenn. May.

P. Tæda L. Loblolly or Old-Field Pine. Low ground; in sandy soil. Frequent southeast of Cleveland, East Tenn.

Abies Fraseri Pursh. Summit of Roane Mt. J. W. Chickering.

Tsuga Canadensis Carrière. (*Abies Canadensis* Michx.) Along water-courses in the mountainous parts of East Tenn.

Taxodium distichum Rich. Along the western course of the Tennessee river; on Mississippi and its affluents; in West Tenn.; cypress swamps.

EQUISETACEÆ.

Equisetum arvense L. Moist meadows, East Tenn.; Cave Spring, Roane Co.

E. robustum Brown. Sandy banks of Mississippi, West Tenn.

FILICES.

Polypodium vulgare L. Cumberland and Alleghany Mts.

P. incanum Swartz. Throughout the State, on rocks and bark of trees.

Cheilanthes vestita Swartz. Bluffs of Cumberland river and Mill creek; also in the cedar glades on the ground and in the mountains of East Tenn.

C. Alabamensis Kunze. Bluffs on Cumberland and Mill creek, near Nashville; Knoxville, East Tenn.

C. tomentosa Link. Bluffs on Ocoe river at Parksville, East Tenn.

Pellæa atropurpurea Link. Dry rocks over the State.

Pteris aquilina L. Siliceous soil, dry hills throughout.

Adiantum pedatum L. Rich, moist ground throughout.

Woodwardia angustifolia Smith. Swamps over the State.

Scolopendrium vulgare Smith. Near New Pittsburgh, Cumberland Mts. Not found by myself.

Camptosurus rhizophyllus Link. Common.

Asplenium parvulum Mart. & Jordan. Shady cliffs over the State.

A. Bradleyi D. C. Eaton. Summit of Lookout Mt., East Tenn.

A. pinnatifidum Nutt. Cumberland plateau, Lookout Mt.

A. Trichomanes L, Siliceous rocks on Ocoe river and over the high mountains of East Tenn.

A. ebeneum Ait. Very common.

A. montanum Willd. Cumberland and Alleghany Mts.

A. Ruta-muraria L. Mountains near Cowan, on sandstone; on limestone rocks near Nashville. Frequent.

A. augustifolium Michx. Rich, damp woodlands. Frequent.

A. thelypteroides Michx. With the former. Highlands.

A. Filix-fœmina Bernh. Common throughout.
· Var *angustum* (*Aspidium angustum* Willd.). With fronds linear-lanceolate. In swamps, West Tenn.

Phegopteris hexagonoptera Fée. Highlands of Middle Tenn.

P. polypodioides Fée. Mountains of East Tenn.

Aspidium acrostichoides Swarz. Throughout.

A. Novæboracense Swarz. Moist woods in the barrens of Middle Tenn.

A. spinulosum Swarz., var. intermedium. Wolf creek, Cocke Co. Summit of Clingman-Dom.

Var. *dilatatum* Gray. Mountains of East Tenn.

A. Goldianum Hook. Oak barrens, Tullahoma.

A. Filix-mas. Swarz. Reported, not yet found by myself.

A. marginale Swarz. Common in the Cumberland and Alleghanies.

Onoclea sensibilis L. Edge of river swamps. Common.
Var. *obtusilobata* Torr. Williamson Co.

Cystopteris fragilis Bernh. From the mountains to the Mississippi.

C. bulbifera Bernh. Cliffs on Richland creek, near Nashville. Cumberland Mts.

Woodsia obtusa Torr. Throughout.

Dicksonia punctilobula Kunze. High mountains of East Tenn.

Trichomanes radicans Swarz. Underneath wet, shelving rocks. Sewanee.

Lygodium palmatum Swarz.   Cumberland Mts.   Very local.
Rockland Station.

Osmunda regalis L.   Swamps, highlands, barrens and mountains.

O. Claytoniana L.   Cumberland and Alleghany Mts.

O. cinnamomea L.   Swamps, throughout the State.

OPHIOGLOSSIACEÆ.

Botrychium ternatum Swarz.   Over the State.

B. Virginicum Swarz.   Over the State.

Ophioglossum vulgatum L.   Cedar glades, at Lavergne.

LYCOPODIACEÆ.

Lycopodium Selago L.   Roane Mt.   Prof. Chickering.

L. dendroideum Michx.   Wolf creek, Cocke Co. ; near Cranberry mines, Johnson Co.

L. complanatum L.   Cumberland Mts., Sewanee.

Selaginella rupestris Sprengel.   Dry rocks, along Ocoe river,
East Tenn.

S. apus Sprengel.   Common, throughout.

Isœtes Butleri, var. *immaculata* Engelm.   Cedar glades near
Lavergne.   In moist places.

HYDROPTERIDES.

Azolla Caroliniana Willd.   Swamps near Johnsonville, West
Tenn.

---

* *Isotes Butleri* Engelm., var. *immaculata* Engelm.   Dioiceous, with a
subglobose trunk, bright green, rather firm leaf, sometimes as many as
sixty, six to nine inches long ; sporangium without spots ; macrospores
0·40-0·56 mm. in diameter ; microspores 0 029-0·031 mm. long, spinulose.
On limestone flats ; in damp places in the cedar barrens.   One mile southeast from the railroad station.   June.

SUMMARY.

Number.	ORDERS.	Genera.	Species.	Varieties.	Species and Varieties.	Introduced Plants.	Woody Plants.	Trees.	Nashville Flora.
1	Ranunculaceæ	16	40	3	43	2	1	...	32
2	Calycanthaceæ	1	2	..	2	...	2	...	1
3	Magnoliaceæ	2	5	...	5	1	...	5	2
4	Anonaceæ	1	1	...	1	...	...	1	1
5	Menispermeæ	3	3	...	3	...	1	...	3
6	Berberideæ	4	4	...	4	...	...	...	3
7	Nymphæaceæ	5	5	...	5	...	...	...	3
8	Papaveraceæ	4	5	...	5	3	...	...	5
9	Fumariaceæ	3	6	...	6	...	...	...	2
10	Cruciferæ	16	34	3	37	10	...	...	30
11	Capparidaceæ	2	2	...	2	...	...	...	1
12	Cistaceæ	2	6	...	6	...	...	...	5
13	Violaceæ	2	12	4	16	1	...	...	12
14	Polygalaceæ	1	8	1	9	...	...	...	4
15	Caryophyllaceæ	7	21	...	21	5	...	...	16
16	Paronychieæ	2	3	...	3	...	...	...	2
17	Portulaccaceæ	3	5	...	5	2	...	...	4
18	Hypericaceæ	3	22	2	24	1	6	...	13
19	Ternstrœmiaceæ	1	1	...	1	...	1	...	...
20	Malvaceæ	7	10	...	10	4	...	...	9
21	Tiliaceæ	1	2	...	2	...	...	2	1
22	Linaceæ	1	4	...	4	1	...	...	3
23	Geraniaceæ	4	8	...	8	...	...	...	6
24	Rutaceæ	2	2	...	2	...	2	2	2
25	Simarubeæ	1	1	...	1	1	...	1	1
26	Meliaceæ	2	2	...	2	2	...	2	2
27	Ilicineæ	2	5	...	5	...	3	2	1
28	Celastrineæ	2	3	...	3	...	2	1	2
29	Rhamnaceæ	3	4	...	4	...	2	2	3
30	Vitaceæ	2	9	...	9	...	8	...	5
31	Sapindaceæ	5	11	1	12	1	1	8	9
32	Anacardiaceæ	1	6	...	6	...	3	3	5
33	Leguminosæ	35	85	3	88	9	3	6	72
34	Rosaceæ	14	53	7	60	10	13	16	47
35	Saxifragaceæ	12	19	1	20	...	8	...	7
36	Crassulaceæ	3	7	..	7	...	...	...	3
37	Hamameliaceæ	2	2	...	2	..	1	1	2
38	Halorageæ	2	3	...	3	...	...	...	3
39	Melastomaceæ	1	2	...	2	...	...	...	1
40	Lythraceæ	6	6	...	6	...	...	...	4
41	Onagraceæ	6	19	2	21	...	...	...	13

SUMMARY—Continued.

Number.	ORDERS.	Genera.	Species.	Varieties.	Species and Varieties.	Introduced Plants.	Woody Plants.	Trees.	Nashville Flora.
42	Passifloreæ	1	2	...	2	...	...	...	2
43	Cucurbitaceæ	4	4	...	4	...	...	...	3
44	Cactaceæ	1	1	...	1	...	...	...	1
45	Ficoideæ	1	1	...	1	...	...	...	1
46	Umbelliferæ	22	29	...	29	4	...	...	24
47	Araliaceæ	1	5	...	5	...	...	...	3
48	Cornaceæ	2	8	...	8	...	3	5	6
49	Caprifoliaceæ	6	12	...	12	...	7	1	6
50	Rubiaceæ	7	20	2	22	...	1	...	16
51	Valerianeæ	1	2	2	4	...	...	...	3
52	Dipsaceæ	1	1	...	1	...	...	...	1
53	Compositæ	64	202	17	219	17	...	...	173
54	Lobeliaceæ	1	8	...	8	...	...	...	7
55	Campanulaceæ	2	4	...	4	...	...	...	2
56	Ericaceæ	16	32	1	33	...	18	5	9
57	Primulaceæ	5	9	1	10	3	1	...	1
58	Sapotaceæ	1	1	...	1	...	1	...	1
59	Ebenaceæ	1	1	...	1	...	...	1	1
60	Styraceæ	1	1	...	1	...	...	1	...
61	Oleaceæ	3	7	1	8	1	1	7	8
62	Apocyneæ	3	4	...	4	1	...	...	4
63	Asclepiadeæ	5	16	1	17	...	...	...	14
64	Loganiaceæ	3	3	...	3	...	1	...	1
65	Gentianeæ	5	10	...	10	...	...	...	7
66	Polemoniaceæ	3	8	1	9	...	...	...	8
67	Hydrophyllaceæ	4	9	1	10	...	...	...	7
68	Boragineæ	8	15	...	15	3	...	...	12
69	Convolvulaceæ	4	17	...	17	4	...	...	14
70	Solaneæ	6	13	...	13	5	...	..	12
71	Scrophulariaceæ	20	36	2	38	4	...	...	30
72	Orobanchaceæ	3	3	...	3	...	...	...	3
73	Lentibulariaceæ	1	2	...	2	...	...	...	...
74	Bignoniaceæ	3	3	...	3	...	...	1	3
75	Pedaliaceæ	1	1	...	1	...	...	...	1
76	Acanthaceæ	4	5	...	5	...	...	...	4
77	Verbenaceæ	4	10	...	10	1	...	...	7
78	Labiatæ	26	59	5	64	12	...	...	54
79	Plantagineæ	1	8	...	8	2	...	...	7
80	Amarantaceæ	3	6	1	7	5	...	...	7
81	Chenopodiaceæ	1	7	1	8	5	...	...	6
82	Phytolaccaceæ	1	1	...	1	...	...	...	1
83	Polygoneæ	4	23	...	23	7	...	...	21
84	Podostemaceæ	1	1	...	1	...	...	...	...
85	Aristolochiaceæ	2	6	...	6	...	2	...	3
86	Nyctagineæ	1	2	...	2	...	...	...	1
87	Saurureæ	1	1	...	1	...	...	...	1

SUMMARY—Continued.

Number.	ORDERS.	Genera.	Species.	Varieties.	Species and Varieties.	Introduced Plants.	Woody Plants.	Trees.	Nashville Flora.
88	Lauraceœ	2	2	...	2	...	1	1	2
89	Thymeleaceæ	1	1	...	1	...	1	...	...
90	Santalaceæ	3	3	...	3	...	...	3	...
91	Loranthaceæ	1	1	...	1	...	...	...	1
92	Euphorbiaceæ	9	24	1	'25	1	...	...	21
93	Urticaceæ	13	18	1	19	3	...	10	16
94	Platanaceæ	1	1	...	1	...	...	1	1
95	Juglandaceæ	2	9	...	9	...	...	9	8
96	Cupuliferæ	6	25	...	25	...	3	22	21
97	Betulaceæ	2	6	...	6	...	1	5	2
98	Salicineæ	2	14	...	14	5	7	7	9
99	Callitrichineæ	1	2	...	2	...	...	...	2
100	Ceratophyllaceæ	1	1	...	1	...	...	...	...
101	Araceæ	4	6	...	6	...	...	...	4
102	Lemnaceæ	3	6	...	6	...	...	...	4
103	Typhaceæ	3	5	...	5	...	...	...	4
104	Naiadeæ	2	7	...	7	...	...	...	3
105	Alismaceæ	2	4	1	5	...	...	...	4
106	Hydrocharideæ	2	2	...	2	...	...	...	1
107	Orchidaceæ	11	28	...	28	...	...	...	15
108	Amaryllidaceæ	3	3	...	3	...	...	...	3
109	Hœmadoraceæ	1	1	...	1	...	...	...	1
110	Iridaceæ	3	9	...	9	1	...	...	6
111	Dioscoreæ	1	1	...	1	...	...	...	1
112	Smilaceæ	1	7	...	7	...	...	...	7
113	Liliaceæ	29	45	2	47	6	...	...	30
114	Juncaceæ	2	16	1	17	...	...	...	10
115	Pontederiaceæ	3	4	...	4	...	...	...	3
116	Commelynaceæ	2	5	1	6	...	...	...	6
117	Xyridaceæ	1	2	...	2	...	...	...	1
118	Cyperaceæ	12	85	6	91	...	...	...	65
119	Gramineæ	53	134	23	157	19	...	...	133
120	Coniferæ	5	10	...	10	...	...	10	2
121	Equisetaceæ	1	2	...	2	...	...	...	...
122	Filices	18	39	3	42	...	...	...	24
123	Ophisglossiaceæ	2	3	...	3	...	...	...	3
124	Lycopodiaceæ	3	6	...	6	...	...	...	2
125	Hydropterides	1	1	...	1	...	...	...	...

DIVISIONS.

GROUPS.	Orders.	Genera	Species.	Varieties.	Species and Varieties.	Introduced.	Woody Plants.	Trees.	Nashville Flora.
Polypetalæ............	48	224	499	27	526	57	60	57	380
Gamopetalæ...........	31	213	522	34	556	53	23	16	416
Dichlamideæ........	79	437	1,021	61	1,082	110	83	73	796
Monochlamideæ.....	21	58	154	4	158	26	15	58	123
Dicotyledones	100	495	1,175	65	1,240	136	98	131	919
Monocotylea..........	19	138	370	34	404	26	...	...	301
Gymnospermæ.......	1	5	10	...	10	...	...	10	2
Phænogamia	120	638	1,555	99	1,654	162	98	141	1,222
Vascular Cryptog...	5	25	51	3	54	...	...	...	29
Total.............	125	663	1,606	102	1,708	162	98	141	1,251

The total of species and varieties, presently known:

    For the State...1,708
    Nashville Flora 1,251

The actual number of species and varieties for the State comes perhaps near 2,000.

CORRIGENDA.

The remoteness of the author from the printer has been a great inconveni-ence in revising and correcting proof sheets. For minor inadvertencies we ask the indulgence of the reader, and correct only the more offensive or misleading ones.

Page 4, line 22. For Sullivant's read Sillimans.
Page 9, line 20. For Triestum read Trisetum.
Page 9, line 27. For serppyllifolia read serpyllifolia.
Page 16, line 53. For racmeosus read racemosus.
Page 19, and follow. For Virgianiana read Virginiana.
Page 21, line 16. For Chilhouewe read Chilhowee.
Page 36, line 13. For stuvei read Stuvei.
Page 37, line 35. For marsh read marshy.
Page 42, line 15. For Rhadiola read Rhodiola.
Page 47, line 19. For Sempervirens read sempervirens.
Page 47, line 25. For Benkley read Buckley.
Page 52, line 19. For Sindl. read Lindl.
Page 53, line 22. For Plinhea read Pluchea.
Page 53, line 24. For Beyrishii read Beyrichii.
Page 53, after E. strigosus transport from below Var. Beyrichii Torr. & Gray.
Page 58, line 3. For obovutus read obovatus.
Page 66, line 15. For Tenn. read Tunnel
Page 77, line 18. For Podosteman read Podostemon.
Page 79, after Cannabis sativa L., dele. Not observed indigenous.
Page 84, line 9. For ophiaglossoides read ophioglossoides.
Page 91, line 6. For polyphillus read polyphyllus.
Page 95, line 12. For dichotomacrum read dichotomum.